The Institution of Structural Engineers

Subsidence

Authors

J C W Patch	BSc FICE (Technical Consultant ASUC) *Chair*
A Boobier	BEng (Hons) CEng FICE
O Brooker	BEng CEng MICE MIStructE MCS (Structronix)
A Bussey	BSc (Hons) DipArchCons FRICS C.Build E MCABE (Gateley Smithers Purslow)
G Carter	BEng (Hons) CEng MICE DipCII (Claims)
B N Deakin	BSc (Hons) CEng MICE FCILA (RSA Insurance)
S Dee Shapland	BSc (Hons) MRICS C.Build E FCABE (MD Group)
T Farewell	PhD MSc BSc (Hons) M.I.SoilSci SFHEA (MapleSky Ltd)
S James	RIBA FICFor (Hon) (Trees and Design Action Group)
D Johnson	BEng (Hons) FGS (Roger Bullivant)
R King	BSc Hons (Falcon Structural Repairs Ltd)
A M Lee	BSc MSc CEng MIMMM CGeol FGS (Geobear Global)
R Mace	IEng IMIStructE (Shire Structures)
M MacQueen	BSc CBiol MRSB MICFor CEnv MAE MEWI (Property Risk Inspection Limited)
P Noble	BTech (Hons) CEng MIStructE (Premier Guarantee)
S Odunmbaku	MSc CEng MIStructE (NHBC)
A Parkinson	MRICS MIRPM (RICS)
T Pither	CEng MICE ACILA AaPS MAPM (Subtilis Consulting Ltd)
Z Rose-Higgins	MEng PhD (Coal Authority)
M Shum	BSc (Hons) C.Build E MCABE (LABC)
J C Smith	MICFor (Forestry Commission)
G Strong	BSc (Hons) FRICS FCIArb C.Build E FCABE FCILA FUEDI-ELAE (RICS)
A D Thompson	MScEDM(Open) CEnv FRICS FCInstCES (Anglia Ruskin University/RICS)
R Withers	(ASUC)

Publishing

L Baldwin	BA (Hons) DipPub (The Institution of Structural Engineers)
R Thomas	BA (Hons) MCLIP (The Institution of Structural Engineers)

Published by The Institution of Structural Engineers
International HQ, 47–58 Bastwick Street, London EC1V 3PS, United Kingdom
T: +44(0)20 7235 4535
E: mail@istructe.org
W: www.istructe.org

Typeset by Riverside Publishing Solutions Ltd
Printed by Cambridge Scholars Publishing

First published (version 1.0) November 2023
978-1-906335-64-9 (print)
978-1-906335-65-6 (pdf)

Acknowledgements

The Institution wishes to thank: Brian Clancy, Laurance Donnelly, Tony Boobier and Owen Brooker for their help in reviewing the material; Moniteye Ltd for their assistance with various images used in the publication, and The Forestry Commission for their generous support in the production of this guidance.

The Institution would also like to thank Matthew Shum for his work in creating:
Figures 2.1–2.4, 5.2, 5.5, 5.8, 5.9, 6.10, 6.11, 6.16–6.18, 7.5, 8.5, 9.3, 9.4, 9.6, 9.9–9.14, 9.17–9.19, 9.22 and D4.

Permission to reproduce the following has been obtained, courtesy of these individuals/organisations:
Figures 1.1 and 5.6: ©John Patch
Figure 1.2: ©ABI
Figure 5.1: ©Dan Gregory/Warnham Landfill Site. Licenced under Creative Commons Attribution 2.0 Generic licence
Figures 5.4, 5.10–5.13, 5.15 and 5.18–5.22 and template Certificate of Structural Adequacy (Appendix B): ©Owen Brooker
Figure 5.16: ©Smithers Purslow
Figure 6.1: Contains British Geological Survey materials ©UKRI 2023
Figures 6.2–6.5, 6.7–6.9 and 6.12: ©Subtilis Consulting Ltd
Figures 6.6, 9.1 and 9.2, 9.5, 9.7, 9.8, 9.15, 9.16 and 9.23–9.27: ©Falcon Structural Repairs
Figures 6.11, 6.13 and 6.18: ©Steven Odunmbaku
Figure 6.14: ©NHBC
Figures 7.1 and 7.2: ©Biddle PG. *Tree root damage to buildings. Vol 1: Causes, diagnosis and remedy*; *Vol 2: Patterns of soil drying in proximity to trees in clay soils*. Wantage: Willowmead Publishing Ltd, 1998
Figures 7.3, 7.4 and 7.6–7.9: ©Shire Structures Ltd
Figure 8.1: ©John Wiley & Sons Ltd
Figures 8.2–8.4: ©London Borough of Camden
Figures 9.20 and 9.21: ©Geobear
Figure 11.1: ©Timothy Farewell (using Met Office and UKCP18 data)
Figure 11.2: ©Timothy Farewell (modelled using data from HadUK-Grid and UKCP18)
Figure D1: Arup Geotechnics. *Review of mining instability in Great Britain.* London: HMSO, 1991 (Crown Copyright)
Figures D2 and D3: ©Littlejohn GS. 'Structures liable to the effects of mining subsidence'. In Henry FDC. ed. *The design and construction of engineering foundations* (2nd edition). London: Chapman & Hall, 1987
Figure D4: ©Matthew Shum (derived/adapted from an image owned by Cornwall Mining Services Ltd)
Specimen underpinning guarantee (Appendix F): ©Association of Specialist Underpinning Contractors (ASUC)

Tables 6.1 and 6.2: Credit: Accuris (from ©BRE. *Assessment of damage in low-rise buildings with particular reference to progressive foundation movement (BRE Digest 251)*). Rev ed. Watford: BRE, 1995

Contents

Foreword

Our health, both physical and mental, is the most important factor that affects our daily lives, but the health of the buildings and other structures we design, construct, occupy and (hopefully eventually) reuse, also has a crucial role to play in our own wellbeing. Structural integrity is one of the most important aspects of structural health — and the most likely cause of structural ill health in the UK, is subsidence.

Just as there have been rapid advances in the medical, psychiatric and psychological fields, *Subsidence* has been prepared by a task group of leading subject experts (from engineers and arboriculturists to legal and insurance professionals) to ensure that the most up to date guidance is available to anyone affected by the problem. Subsidence can affect all types of buildings and structures and the range of problems can span minor cracking to total collapse. One thing is for certain… as with human health it is necessary to continuously improve how we identify the problems, diagnose the conditions and rectify the failures. *Subsidence* has been created to do just that.

The IStructE has published guidance on this subject in the past (the first edition of *Subsidence of low-rise buildings* was published in 1994 and a second edition followed in 2000) providing an all-embracing guide to the subject of subsidence rectification. This new publication draws on significant amounts of development in the intervening 23 years, in areas such as data capture and trending analysis, improvements in diagnostic and investigative methods, and additional rectification processes.

Subsidence certainly draws on the experience of the earlier guidance; namely the gathering together of a formidable team of specialists and experts to create a new publication which covers the entire subject of subsidence — from the discovery of a crack to the sign-off of a completed rectification project.

I am immensely grateful to the Chairman of the task groups that produced both editions of the previous guidance, Mr Brian Clancy. His dedication to this subject is enormous, without which the issue would not be so well documented.

I am also grateful to him for choosing me as the Chair for this publication. I was a member of the two previous task groups and it is an honour to have directed this endeavour.

However, the people to whom I am most grateful are the members of the current task group; their passion for the subject, and dedication to producing this book is much appreciated and has involved many hours of discussion, writing, reflection, review and rewriting!

Lastly, the support and practical input of IStructE staff, Lee Baldwin and Rob Thomas has been exemplary. Thank you to all who have created this powerful guide to understanding, and dealing with, subsidence.

John Patch

John Patch
Chair

1 Introduction

1.1 Background

This guidance provides a rational approach to damage and loss in the UK caused by subsidence, heave and landslip, and to the repair of affected buildings. It is an authoritative guide for all those involved in subsidence. Although it focuses solely on low-rise, conventionally constructed buildings (i.e., those up to four storeys) the principles can also be applied to much larger buildings. The guidance is therefore applicable to virtually all domestic properties, except high-rise apartment buildings. The technical principles can also be applied to low-rise buildings used for commercial purposes. The insurance information given is relevant to domestic properties only — and it is recognised that not all buildings are insured.

Subsidence and heave have always affected buildings under certain ground conditions, but in the past, cracks (and other damage resulting from them) were normally attended to by Owners as part of routine maintenance. Very few buildings required major remedial works, although only buildings constructed on very stable hard material, such as rock, are likely to avoid this type of damage completely.

Historically, 'subsidence', 'heave', 'settlement', 'consolidation', 'compaction' and similar terms were used somewhat interchangeably, and this guidance has drawn on the expertise of specialists from various fields to provide some distinction (Chapter 2).

1.2 Historical perspective

1.2.1 Understanding the problem

Insurance cover against subsidence, heave or landslip damage was introduced more than 50 years ago to accommodate Lenders in protecting both their loan security and their customers against catastrophe (Figure. 1.1).

Figure 1.1: Major subsidence damage resulting from collapse of an underground canal tunnel

Since then, cover at times seems to have unintentionally developed to become a form of general building maintenance paid for by Insurers at enormous cost and used for the repair of thousands of properties showing only the slightest signs of subsidence damage (resulting in higher premiums for all). The main problem had been the failure of all involved to anticipate and answer the following questions:

- What is subsidence? Is it different from settlement? What is heave?
- In what circumstances is it reasonable to conclude that a property requires remedial work to overcome the results of subsidence or heave?
- What form should that remedial work take?
- Is it reasonable for the Insured to expect an Insurer to pay for the repair of structurally trivial damage, even if it is technically the result of subsidence, when there is a possibility of the same sort of damage occurring again in the future because of continued slight movement of the property?
- Is it reasonable to expect Insurers to pay for preventative remedial work to a building or part of a building which may not have been damaged by subsidence, heave or landslip, but which might possibly be damaged in the future?

Since the introduction of subsidence insurance cover, there have been several exceptionally dry periods in the UK. During these years, subsidence claims reached peaks which were completely unforeseen (Figure 1.2).

Figure 1.2: **Cost and number of claims in UK, 1975–2021**

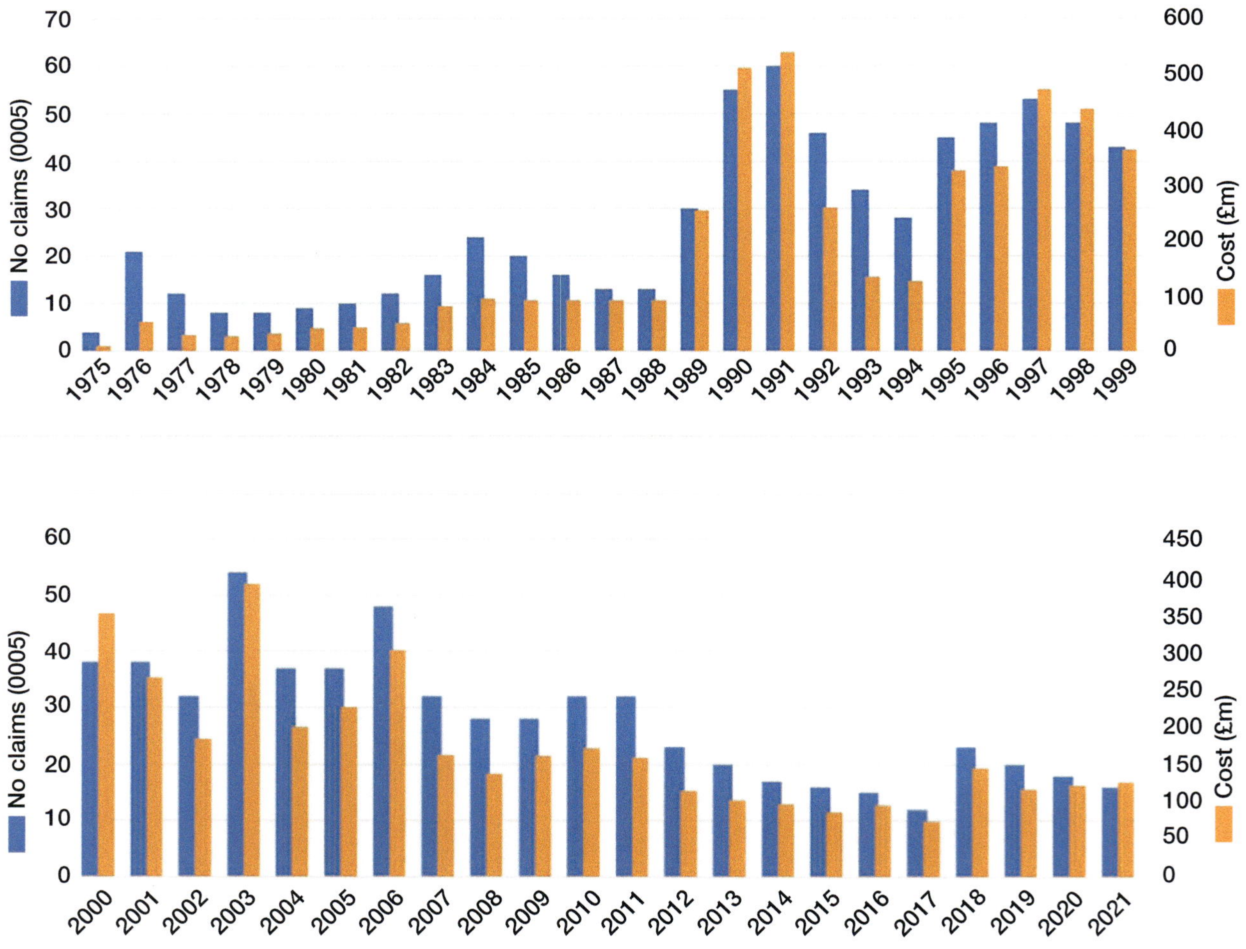

Credit: data from ABI
*cost estimated

Thousands of properties suffered structural movement, cracks appeared in walls and ceilings, and Homeowners were obviously extremely worried for two reasons:

1. Were their homes becoming structurally unsound and therefore dangerous?
2. Were their homes being affected so severely that their value would decrease appreciably?

There was no doubt that in these dry periods, a number of properties suffered quite severe structural movement, to the extent that even engineers had concerns about their structural stability. However, the vast majority suffered only minor damage.

As Homeowners became more aware that their properties had insurance cover for subsidence damage, it was inevitable that claims were going to be made by the Owners against their Insurers even in trivial cases. Insurers also encouraged this by insisting on 'early notification of damage' to ensure mitigating action was taken (Section 4.4.2). Unfortunately, the extent to which a property must be damaged before it technically justifies remedial works had never been defined by Insurers, so the obvious course of action for the Owner was simply to make a claim and have the work attended to at the Insurer's expense. A large policy excess was also introduced to discourage trivial claims. All this has since given rise to problems and misunderstandings where Insurers have difficulty managing the Insured's expectation over the appropriateness of repair and the destinction between that repair and general maintenance.

1.2.2 Reacting to the problem

Unfortunately, when subsidence management went wrong it was likely to have been because the problem was not examined from first principles.

Typically, it was not recognised that building construction that took place during the 1950s through to the 1980s had used less-flexible materials. For example, hard cement mortar was introduced in place of traditional lime mortar for brickwork, making buildings more brittle. As a result, buildings from this period tend to show signs of structural movement more dramatically than older, more-flexible buildings, and even minor movement may be sufficient to cause some visible damage.

Consequently, because individual Insurers were having to meet many thousands of subsidence claims, they became much more reluctant to provide cover for any property with signs of subsidence damage or structural movement. In addition, since the early 1970s, public expectation of the structural performance (e.g., tolerance to minor cracking) of buildings, and the quality of service from their professional advisers, had increased dramatically. As a consequence, a number of Court cases brought against professional advisers proved successful — with the inevitable result that these advisers became more conservative with the advice they gave. Surveyors and Engineers, when asked to value or assess properties suffering from suspected subsidence damage or structural movement, found it easier to recommend remedial works rather than appraise the properties more objectively, even when the damage was relatively trivial. By adopting this approach, they reduced the risk of legal actions against them for negligence.

Finally and inevitably, many structural damage problems are identified when properties are advertised for sale. Tight timescales placed considerable pressure on all parties to deal with any problem promptly and to a standard which left no doubt as to the future stability of the property. Indeed, the major drought of 1975–6 caused such widespread damage that for many years, Lenders were cautious about lending on subsidence- and heave-damaged properties even after they had been repaired.

Originally, Insurers and their Experts adopted an over-cautious approach, which meant that many properties suffering from relatively minor damage were underpinned. In the vast majority of cases this has proved to be technically unnecessary. The problem could have been dealt with much more economically and objectively by taking time to identify and deal with the actual cause of damage. However, the demand to repair properties promptly generally meant that this option was not pursued.

Historically, dry summers have resulted in a large industry geared solely to the underpinning of domestic properties suffering from ground movement.

Vast sums have been spent on underpinning works and general maintenance repairs — and most Civil and Structural Engineers would consider this to be substantially unjustified.

1.2.3 The IStructE initiative

The first edition of *Subsidence of low-rise buildings: A guide for professionals and property owners*[1] was published in 1994. It was a huge collaborative effort that drew on the expertise of a wide-ranging task group, that considered the views of more than 80 contractors who had attended an Institution seminar on the subject in 1992.

Although the first edition was successful in focusing diverse professional interests towards common guidelines for dealing with subsidence, heave and landslip, by 1998 there had been numerous developments in procedures and legislation which highlighted the need for a second edition. In particular, the first edition was not thought to give sufficient guidance on trees. Authors of the second edition (published in 2000) included arboriculturists, not only to help amend and update the guidance generally, but to write an entire chapter devoted to trees and tree management[2].

Finally, although the first edition was intended to reassure Valuers and Homeowners that minor cracking was of little consequence structurally and should not affect property values, over-caution on the part of Valuers and property Owners remained. Insurance claims were increasingly being made as a result of this type of damage even though most Experts regarded the damage as being the natural consequence of 'normal movement' in low-rise buildings. As a result, the second edition of the guidance defined this type of damage and movement more precisely, with a view to bringing a greater degree of understanding and realism to the assessment of damage to low-rise buildings.

1.3 Today's perspective

Subsidence claims and costs since 2000 have continued to be significant — with 'surge years' (Section 1.4) in 2003 and 2006, and further significant activity also seen in 2010, 2011 and 2022.

Although the causes of ground movement remain unchanged (and readers should gain an understanding of the mechanics behind this causation, as well as other defects commonly confused with movement) methods to mitigate damage (Appendix A), repair buildings and restore stability (and the timescales associated with these undertakings) have advanced considerably — with all stakeholders looking to improve the 'claim experience' for Policyholders.

Where previously, Insurers and their Adjusters would predominantly look to engage independent building professionals to review subsidence claims, establish cause and organise suitable repairs, this is now the exception. Most Adjusters have internal teams that focus purely on resolving subsidence matters without independent help, and new claims management firms have been formed that work solely on subsidence matters. The upshot is that claims can be handled more rapidly and cost effectively and in some instances for a fixed, pre-determined outlay. Independent professionals still remain heavily involved in subsidence matters but generally on more complex cases.

Over the last 25 years, new operating models have emerged for the handling of insurance claims. These range from the use of internal staff, usually engineering qualified, to the use of external subsidence management companies. In some cases, the Expert or repair company is subject to a 'term' or 'framework' agreement.

In the case of both Experts and Repairers, formal procurement processes and the adoption of modern supplier management techniques have been used to improve performance, reduce complaints, optimise repair cost and to leverage the buying power of Insurers. This approach aims to improve customer service and reduce fraud, and can provide a sound environment for innovation.

Often services are subcontracted to smaller organisations but this may have led to a loss of market capacity and important expertise which could create issues in times of surge. It is outside the scope of this guide to offer commentary on the pros and cons of these commercial arrangements, other than to say that the principles of fairness and good practice should prevail.

New processes have emerged in the handling of subsidence claims especially through the use, by Insurers, of framework and term agreements for the appointment of Experts, Repairers and subsidence management companies. In some cases the Expert may be a direct employee of the insurance company themselves, and the Repairer directly introduced by (or directly appointed by) the Insurer.

Any Expert or subsidence management company appointed by an Insurer directly, or in the direct employment of an Insurer as part of their staff, should not only to adopt the best practice outlined in this guidance, but also recognise their professional duty to the Homeowner.

Where a Repairer is introduced to the Policyholder, or even in some cases directly appointed by an Insurer or their representative, formal contractual relationships should be properly constructed and be appropriate to the particular process in place.

Where an Expert and/or Repairer has been directly introduced to an unknowing Homeowner by an Insurer, that Insurer owes a special duty in terms of the quality of service provided.

In 2021 the Institution established a contributor group (each member an authority in their particular field) to collectively author this publication, to address the changing landscape and current issues facing the profession. Figure 1.3 presents an overview of how a subsidence claim may be handled.

1.4 A note on 'surge years'

Surge years in the UK are characterised by there being more than 50,000 subsidence claims[3]. The primary cause of a surge is usually an extended period of exceptionally hot, dry weather between June and September. Surges have also been triggered by record-breaking summer temperatures. An extended period of dry weather ahead of summer can increase the amplitude of summer claim numbers, which allows for a degree of prediction[4].

Chapter 11 considers the emergence of improved analytic modelling, which will help Insurers and practitioners anticipate the likelihood of larger numbers of claims, and aid the preparation of their servicing.

While surge years invariably place great financial pressure on Insurers, the impact on the Homeowner as Policyholder should not be underestimated.

Key aspects that need to be considered include:

- Recognition that surge years can, to some degree, be anticipated, and as such be planned for
- Framework agreements should have robust contingency plans for surge and, importantly, should be appropriately stress tested
- Supply chains and processes should be resilient
- Monitoring should not be used as a process for levelling workloads/as a substitute for good practice
- There needs to be recognition of the mental stress on practitioners where large volumes of claims are lodged within a relatively short period
- The use of effective communication plans — to help in the management of customer expectations

Figure 1.3: Overview of subsidence claim

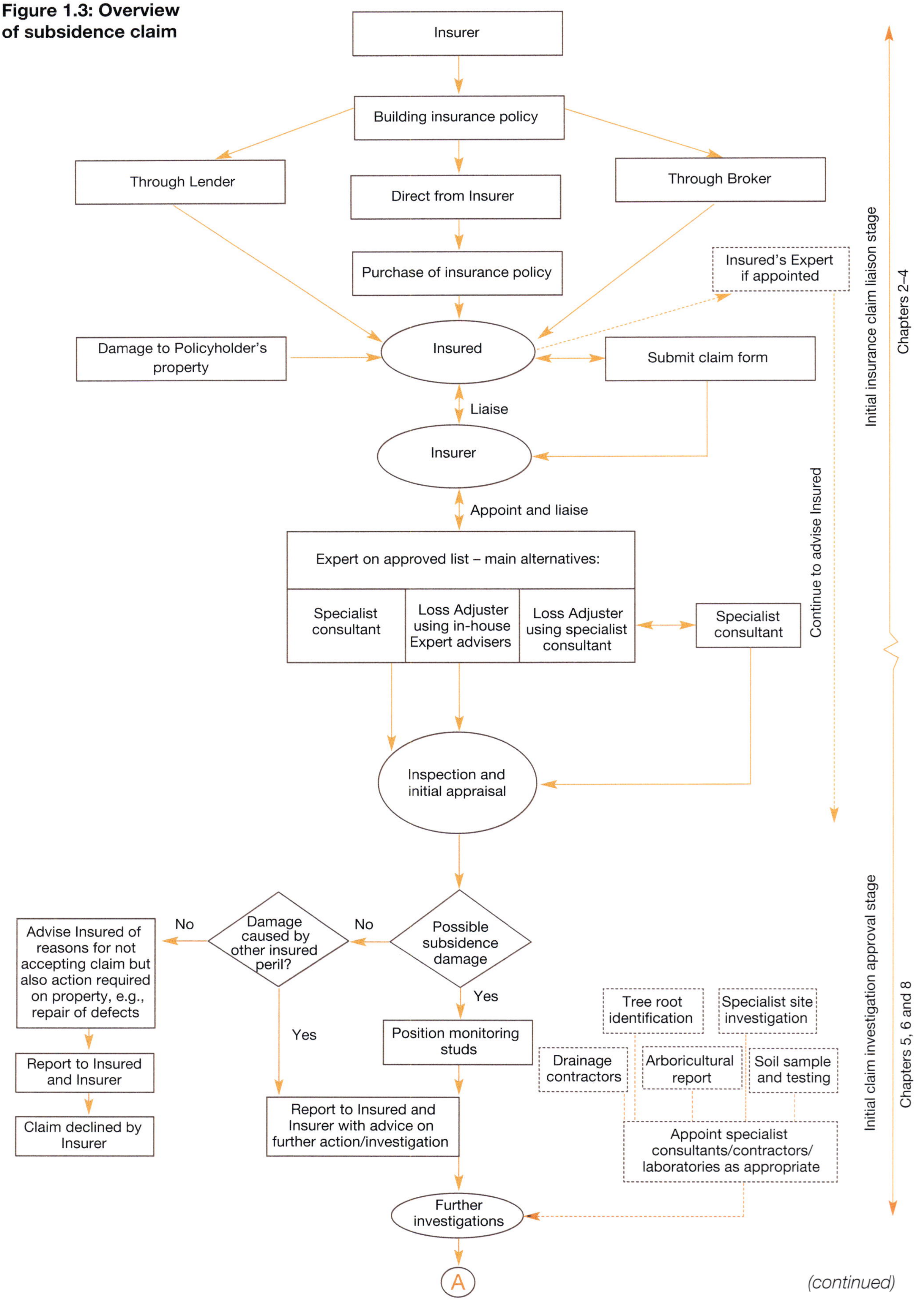

(continued)

Figure 1.3: ***continued***

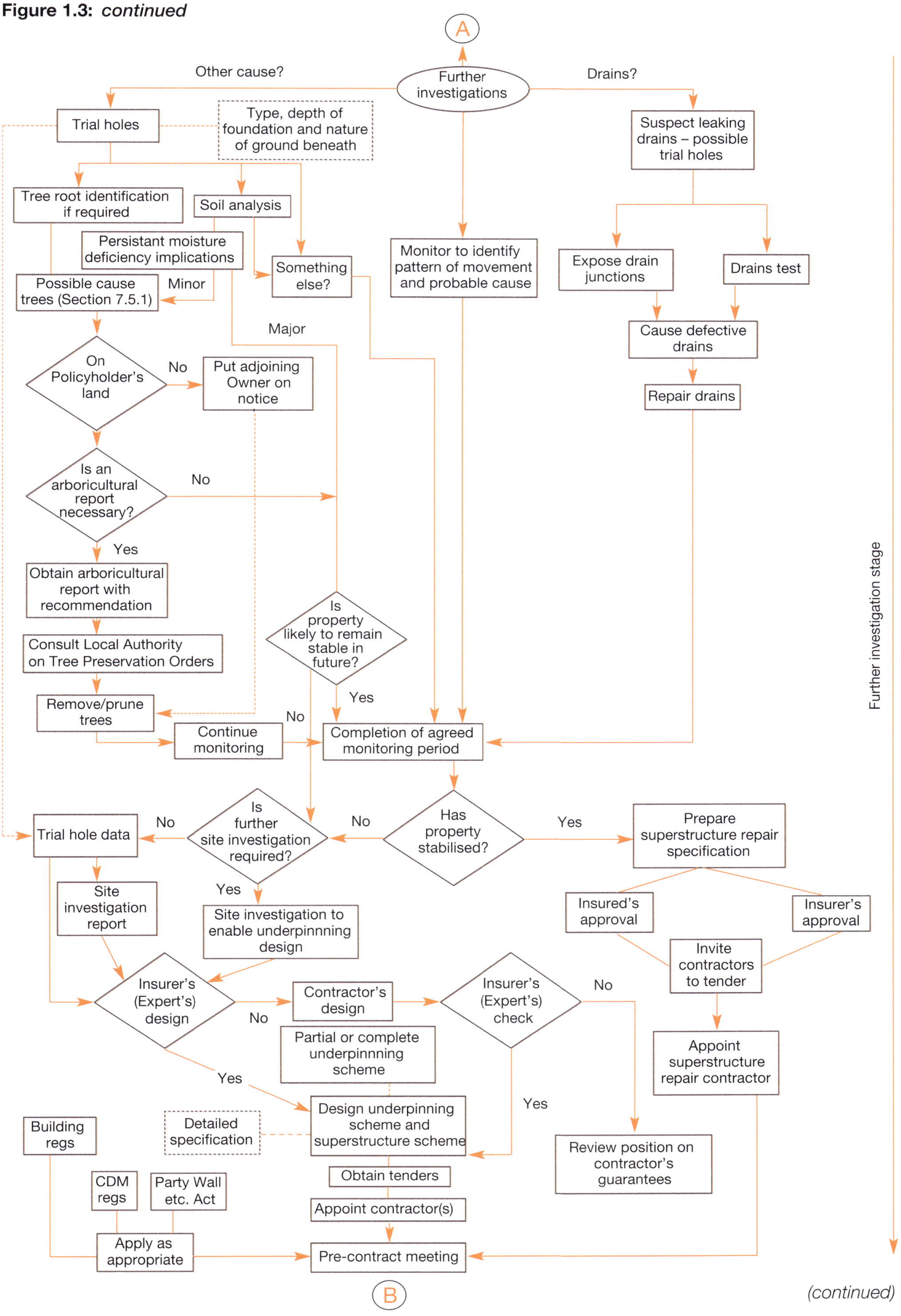

(continued)

Figure 1.3: ***continued***

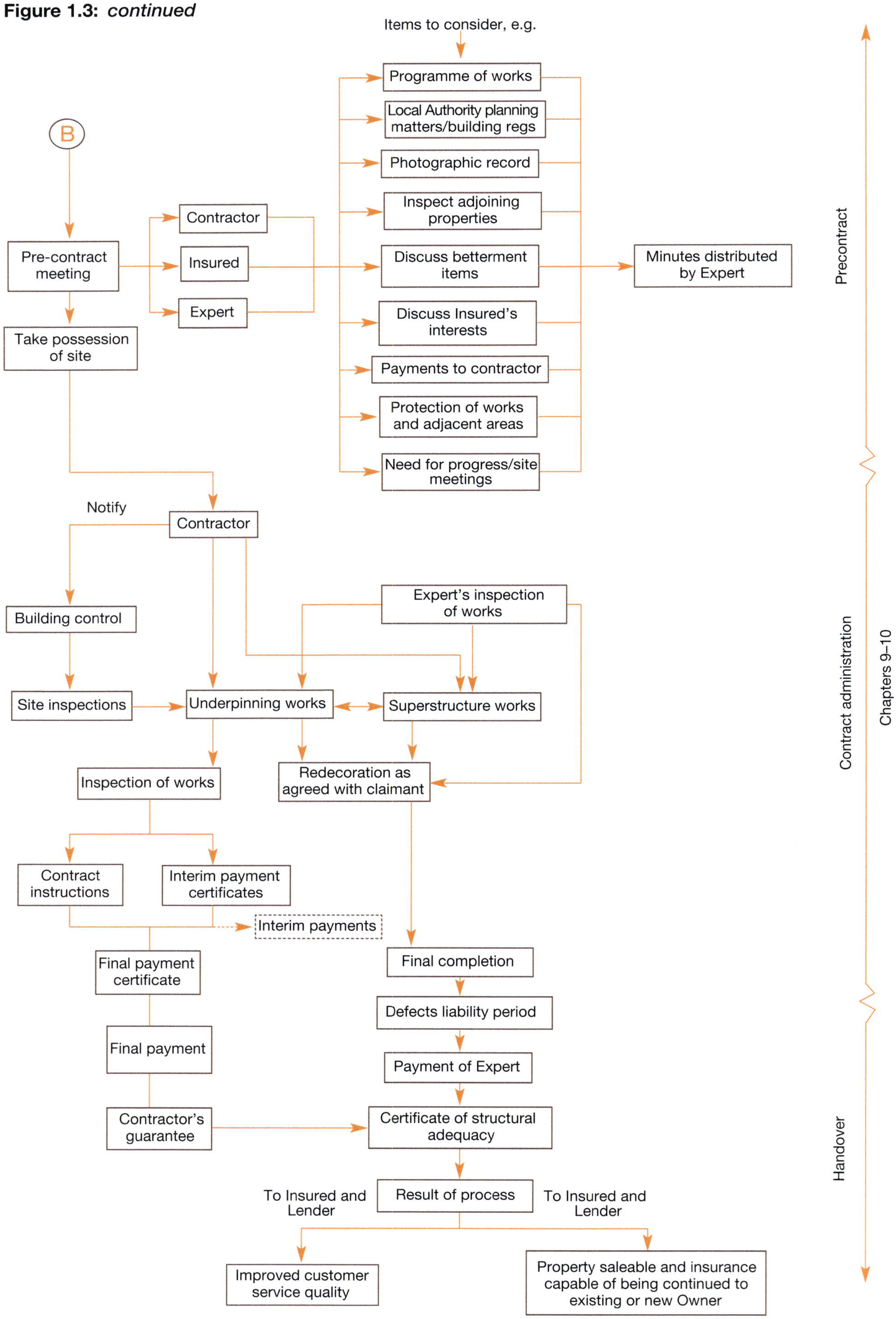

2 Terms relating to building damage†, and a description of the mortgage process

2.1 Terms

2.1.1 Technical terms

The starting point for all technical terms should be the terms and conditions of the insurance policy that applies to the claim being dealt with. It is important to carefully consider these terms to determine the extent of the insurance cover provided. In the event of a claim with no insurance cover, descriptions given here can form a guide. The definitions of 'subsidence' and 'heave' are generally accepted, and there is broad agreement that both terms are distinguishable from 'settlement'.

Settlement

Settlement is commonly considered to be movement within a structure due to the distribution or redistribution of loading and stresses within the various elements of construction. This normally occurs in the early stages of a building's life and may be associated with the initial compaction, consolidation, or movement of the ground beneath the foundations due to the weight of the building. This initial settlement is not normally an 'insured peril' (Section 2.1.2).

Cracking caused by this initial settlement is not normally a continuing problem. Should initial settlement result in major structural damage, this could be dealt with under the terms of a developers' warranty or by the original builder.

Reactivation of settlement movement can subsequently occur if additional loading is imposed onto the property, e.g., first floor extension, loft conversion or reconfiguration of internal walls.

From an insurance perspective, the Financial Ombudsman Service (FOS) defines settlement as:

> "Downward movement as a result of soil being compressed by the weight of a building within 10 years of construction."[5]

This is an arbitrary level and initial settlement can occur after this date. While some insurance policies include their own definition, most do not.

Subsidence

Subsidence is usually associated with volumetric changes to the ground that is supporting the foundations, e.g.,

- Shrinkage of clay soils
- Softening of soils
- Further compaction or consolidation of fill
- Washing away of fines in granular material
- Dissolution of rock (gypsum/chalk)
- Compression of peat, other than that due directly to the self-weight of the building
- Collapse of historic mine workings

† These are not legally binding definitions, and reference must be made to the terms of individual insurance policies.

It should be emphasised that all these volumetric changes can only occur as a direct result of an external factor, e.g.,

- Change in groundwater level
- Leaking drains
- Influence of trees or other vegetation
- Geological features
- Mining
- Compression and decomposition of organic material
- Extremes of climate change
- Adjacent excavations

Subsidence normally requires an external influence to trigger the movement (Figures 2.1 and 2.2).

Figure 2.1: Typical subsidence movement patterns

Figure 2.2: Typical subsidence crack patterns

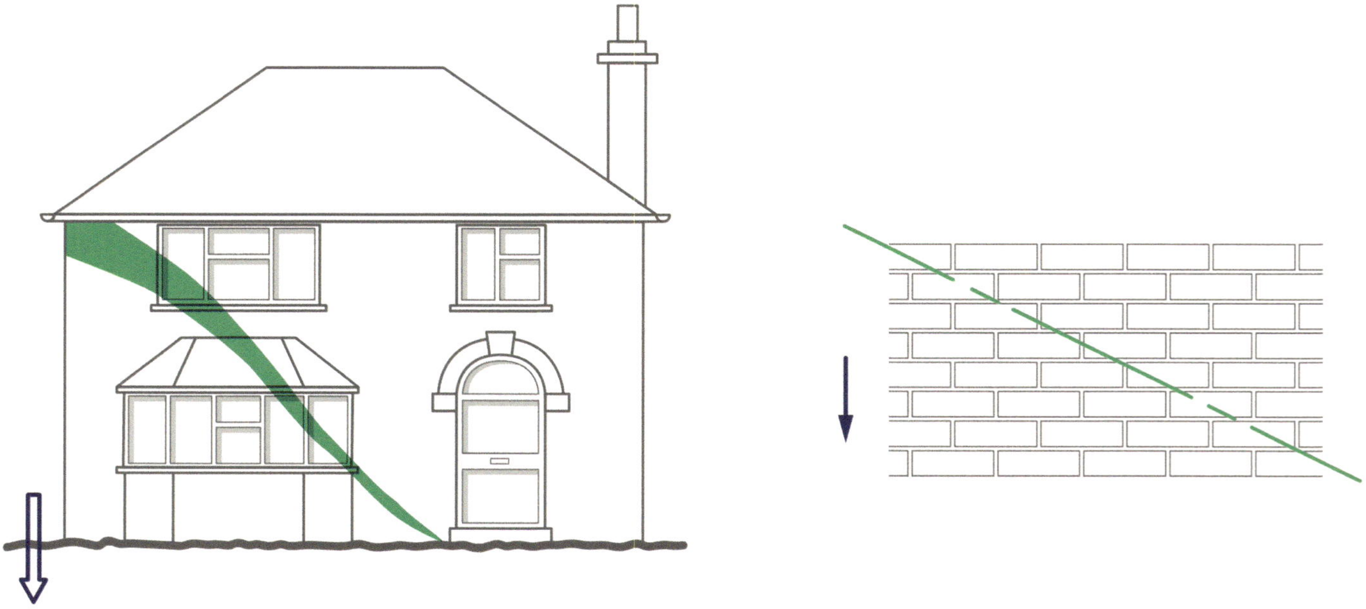

If subsidence is uniform across the foundations, significant damage in the form of cracking to the superstructure is unlikely to occur. However, it may still damage drains. If differential subsidence occurs, remedial measures to stabilise the foundations and to repair damage may be required to bring the movement within acceptable limits.

The principal test to distinguish between subsidence or settlement of the site is to establish whether the downward movement of the ground would have occurred, at least to some extent, with no applied load from the building. If movement of the site would have occurred without the applied load, damage is likely to be a result of subsidence.

There is no universally accepted definition of subsidence in the industry. Insurers will define what they mean by subsidence in the terms and conditions of their policies. The FOS and the Association of British Insurers (ABI) have their own definitions.

The FOS defines clay root subsidence as the situation when:

> *"The ground beneath a building sinks, pulling the property's foundations down with it. It usually occurs when the ground loses moisture and shrinks, which can be caused by prolonged dry spells. It may also be caused by trees and shrubs which can absorb significant volumes of water from the soil."*[5]

The FOS recognises that subsidence can occur in other situations, such as escape of water, collapse of solution features (also known as 'dissolution features'), mining, poor ground and decomposing organic fill.

Heave

Heave is the volumetric expansion of the ground beneath part, or all, of the building. This is normally taken to mean upwards movement of the site caused by expansion or swelling of the subsoil (Figure 2.3). This can result from:

- Moisture increase in clay soils, e.g., from the removal, reduction or death of vegetation
- Removal/reduction of load
- Frost (as water freezes and expands)

Figure 2.3: Typical heave patterns

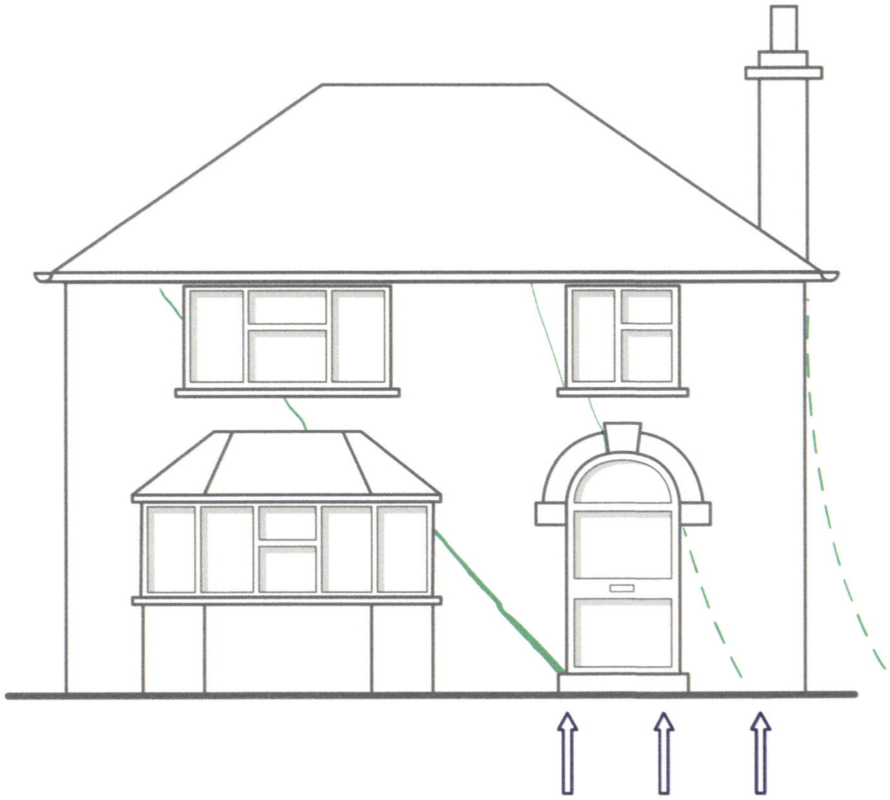

On new build developments, heave protection should be considered when designing foundations on sites with clay soils where extensive tree removal has been completed immediately prior to development.

As with subsidence, there is no universally accepted definition of heave in the industry. The FOS and ABI have their own definitions.

The FOS defines heave as:

> "Upward movement of the ground beneath a building as a result of the soil expanding."[5]

Insurance claims in the UK for damage due to heave are rare.

Recovery/rehydration (of soil)

Recovery/rehydration occurs in clay soil and is the partial reversal of previous subsidence following vegetation management to mitigate the cause of subsidence damage. This is usually not detrimental to the long-term stability of the property and often results in partial or total closure of cracking during the following winter season.

Very occasionally, heave in the form of recovery/rehydration or long-term recovery can only occur if a persistent soil moisture deficit is present and was present at the time of construction of the property. This is where the soil remains in a partially desiccated condition following the process of re-wetting during the winter months. It is more likely that a persistent soil moisture deficit will develop in the presence of mature trees that have a high water demand, but each situation should be considered individually based on site conditions.

Landslip

Landslip is the sudden movement of soil on a slope, or gradual creep of a slope over a period of time. It can involve both downward and lateral movement (Figure 2.4).

Figure 2.4: Landslip

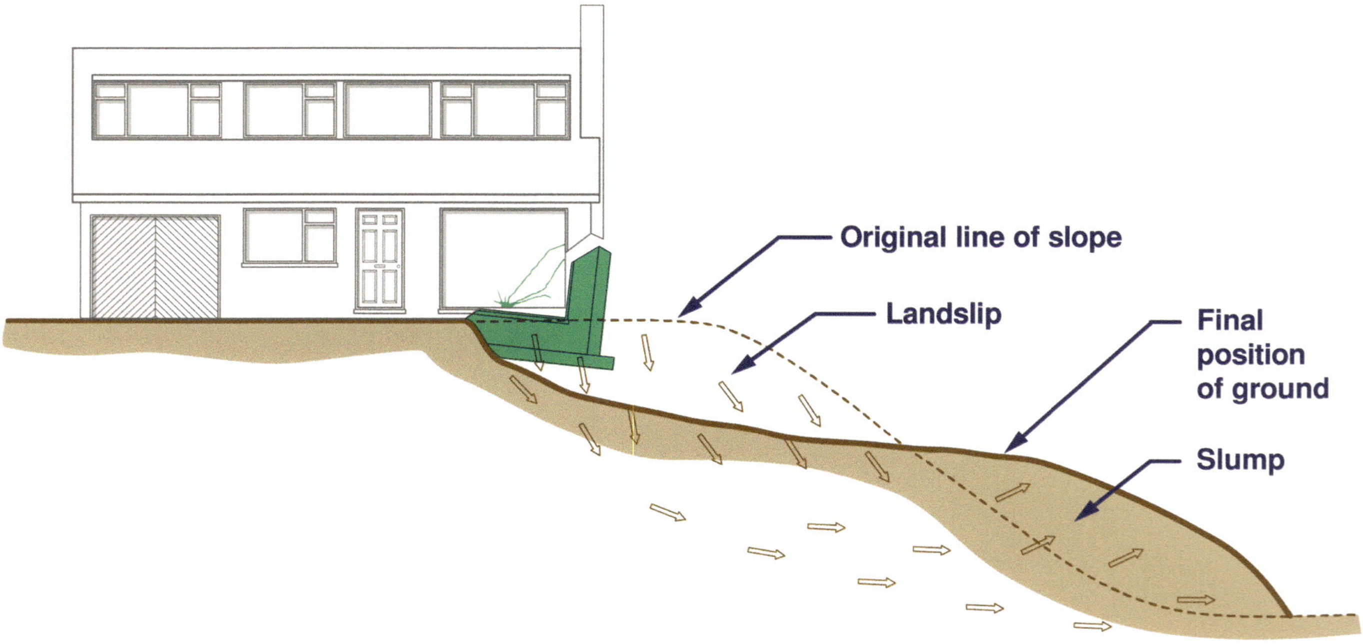

There is no universally accepted definition of landslip in the industry. Insurers will define what they mean by landslip in the terms and conditions of their policies. The FOS and ABI have their own definitions.

The FOS defines landslip as:

> *"Downward movement of sloping ground."*[5]

e.g., natural erosion, excavation at the base of a slope, movement of slope caused by movement of groundwater, leaking drains, stacking of material and fill near the top of a slope, etc.

Insurance claims in the UK for damage due to landslip are rare.

Damage threshold
This is the borderline between damage which is inconsequential and unlikely to be of concern, and that which is likely to be of concern to property Owners generally (and particularly at the point of sale/purchase). For property Owners, the acceptability or otherwise of damage is often a subjective matter. However, professionals who appraise properties need to be more objective, otherwise properties with very minor inconsequential cracks could be incorrectly considered structurally significant or could be the subject of unnecessary insurance claims. One of the key issues is to consider whether the damage is likely to be progressive in the short- or long-term.

There is no universal agreement on what constitutes an acceptable damage threshold, since much will depend on factors such as the property type and construction, soil types and presence of vegetation. Two important issues for consideration are:

1. Is the damage likely to continue/worsen? If not, why has the damage receded/halted?
2. Is the damage of such a serious nature (visually and/or structurally) that intervention can be justified?

The BRE's DG 251 states:

> *"In general, categories 0, 1 and 2 with crack widths up to 5mm can be regarded as 'aesthetic' issues that require only redecoration. Categories 3 and 4 can generally be regarded as 'serviceability' issues, that is, they affect the weather-tightness of the building and the operation of doors and windows. Category 5 presents 'stability' issues likely to require structural intervention."*[6]

DG 251 is used widely in the industry as a way of classifying cracks and determining whether intervention is necessary.

It should be stressed that these comments are a simplification of the assessment needed to properly classify damage to a property. Several factors, including whether the widths of the cracks are increasing with time, can affect the classification.

Tree
A woody plant distinguished from a shrub by its larger size and from having a single, or at least only a small number of, main stems.

Large woody shrub
This term could include large woody shrubs, hedges and mature climbers. It can be difficult to distinguish between a large shrub and a small tree, but in general shrubs tend to produce multiple woody stems that sprout relatively close to the ground, while trees usually have a single trunk.

Climate change
The long-term shift in average weather patterns across the world. Since the mid-1800s humans have contributed to the release of carbon dioxide and other greenhouse gases into the air. This causes global temperatures to rise, resulting in long-term changes to the climate.

Greenhouse gases
A greenhouse gas (GHG) is a gas that absorbs and emits radiant energy within the thermal infrared range, causing the greenhouse effect. Human activities since the beginning of the industrial revolution have increased the atmospheric concentration of carbon dioxide (a GHG). The vast majority of anthropogenic carbon dioxide emissions come from the combustion of fossil fuels, deforestation and other changes in land use.

Anthropogenic emissions
The emission of greenhouse gases into the atmosphere as a result of human activities.

Net zero

The Intergovernmental Panel on Climate Change (IPCC) defines 'net zero emissions' as being achieved when:

> "anthropogenic emissions of greenhouse gases to the atmosphere are balanced by anthropogenic removals over a specific period."[7]

Decarbonisation

The process of reducing the amount of carbon dioxide released into the atmosphere, with the aim of ultimately reaching net zero emissions.

2.1.2 Insurance terms

The insured events (also known as 'insured perils' or 'insured risks') within the scope of this chapter are subsidence, heave and landslip. In establishing whether damage has been caused by one of these insured perils, careful thought needs to be given to whether there has been some other factor, or combination of factors, which could have caused the damage to the building, such as roof spread, lintel failure or thermal movement.

It is also appropriate to consider the terms in this chapter, with the guidance provided in Chapter 4. There will be variance between commercial and residential policies and between Insurers offering those polices. Reference to the terms and conditions of the particular insurance policy should be made in each case to avoid misinterpretation. The definitions contained in the relevant insurance policy terms and conditions should always take priority over the general guidance given here.

The site

In most cases, insurance policies cover subsidence of 'the site' which is generally considered to be the prepared ground on which the building is erected after the foundation trenches have been dug.

Reinstatement cost (sum insured)

The total cost to rebuild the whole of the insured property. The 'whole property' is normally defined as all the structures built on the site, i.e.:

- Main property (e.g., house)
- Detached/attached garage
- Outbuildings
- Swimming pools and tennis courts
- Drives, paths, garden and/or retaining walls, patios, fences

It does not include the value of the ground within the boundaries of the site. For domestic properties, it is usually calculated by multiplying the total area of all floors of the property by a basic cost per unit area, using the Building Cost Information Service (BCIS)[8].

The figure should also include the cost of employing a professional building advisor, e.g., Architect or Surveyor, to assist with the administration of the rebuilding contract and this normally adds to the basic rebuilding sum to make up the full 'sum insured'. The BCIS figures automatically include an amount for professional fees as well as costs incurred in demolition and site clearance where necessary.

The sum insured based on a reinstatement cost assessment is not to be confused with the market value of the property.

Policy conditions

These are both the written policy conditions and the general conditions applicable to the whole policy, and are set out in the policy wording, schedule and endorsements. They set out the terms of the insurance contract between the Insured and Insurer; if they are not complied with, either party may be able to enforce the contract or bring it to an end.

Claims conditions

These conditions, often set out in the 'general conditions' section of a policy, set out what needs to be done by an Insured when a claim is made under the policy. They also explain what rights the Insurer has in dealing with a claim; such as seeking payments from other parties who may have been responsible for the damage.

Exclusions

Exclusions in the insurance policy set out the situations in which insurance cover will not be provided by the Insurer. Again, reference to the terms and conditions of the particular policy should be made.

There are usually specific exclusions which relate to particular sections of the policy. For example, many of the following will probably be excluded from cover provided under a typical subsidence, heave and landslip section of the policy:

- Damage resulting from demolition
- Damage caused by structural repairs or alterations to insured property (note, damage caused by such works being undertaken to an adjacent property may fall under the policy)
- Damage due to defective design or poor workmanship
- Damage due to the use of faulty materials during construction
- Damage to external features (typically paths, patios, driveways, tennis courts, swimming pools, garden/boundary walls, fences, etc.) unless damage has occurred to the main property at the same time by the same cause
- Solid floors unless the external walls are damaged at the same time by the same cause
- Damage due to chemical attack
- Damage due to settlement
- Damage due to settlement of made-up ground

It is important to understand the precise wording of the exclusions in the policy terms and conditions.

Complaints procedures

Each Insurer will have their own complaints procedure which should be clearly set out in their policy terms and should be followed by the customer in the event of a dispute on their claim. Unless the insurance company's complaints procedure is followed, the Insured cannot take a complaint to the Financial Ombudsman Service.

Financial Ombudsman Service

The Financial Ombudsman Service (FOS) helps settle disputes between consumers and UK-based business providing financial services. This includes household insurance companies. The FOS can deal with complaints from consumers about insurance matters. Before the FOS can become involved, the consumer must first give the business that they are unhappy with, the opportunity to look into the complaint itself via that company's complaints service.

Financial Conduct Authority

The Financial Conduct Authority (FCA) is a financial body which operates independently of the UK Government to regulate financial firms providing services to consumers, with the intention of maintaining the integrity of the financial markets in the UK.

Subrogation

A term describing a right held by most insurance carriers to legally pursue a third party that caused an insurance loss to the Insured[9].

Contractor guarantee

Backed by the contractor themselves with no insurance behind it other than (possibly) Professional Indemnity (PI) insurance, should the contractor have such a policy. Conditions of the PI insurance contract will apply.

Contractor Insurance Backed Guarantee (IBG)
Backed by an insurance policy that will benefit the client if the contractor ceases to trade for any reason. The Insurer takes the place of the contractor so there is still a need to prove negligence in order to invoke the guarantee. If negligence is contested, the claim can be risky and expensive.

Warranty
A structural or building warranty is a 'structural defects liability insurance'. Warranty Providers carry out inspections of the property as the build progresses and, on completion, a warranty is provided covering major defects in design, materials or workmanship. In order to bring a claim under a warranty no negligence needs to be proven. However, it must be demonstrated that a claim exists. In practical terms, the first two years following completion of the build resides with the building contractors; if they fail to remedy any damage the policy will take effect. The warranty belongs to the property so there is no need to transfer it at the point of sale. Policy limits are generally contract value plus inflation and alternative accommodation costs.

2.1.3 Terms for relevant professionals

Technical expert adviser (Expert)
Subsidence claims are handled predominantly by Insurers and their technical advisors.

The investigation and rectification of subsidence damage can involve many specialist advisers. However, it is important for the property Owner (and also the Insurer) that a single party should have overall technical and managerial responsibility; from the initial identification of damage through to the issuing of a final report to the Insured and/or a Certificate of Structural Adequacy (Appendix B). Such a technical expert adviser is referred to in this guidance as the 'Expert'. The responsibility of the Expert is to be familiar with all aspects of a subsidence issue.

The Expert is obliged to demonstrate reasonable skill and care when advising on the type of work for which they have been engaged by the client.

The Expert is normally required at three stages:

Stage 1
Initial appraisal (Chapter 6)
Further investigation (Chapter 7)

Stage 2
Mitigation (Chapter 8)

Stage 3
Remedial works (Chapters 8, 9 and 10)

Experts engaged to advise on subsidence, particularly in assessing whether remedial works may or may not be needed, should have detailed knowledge of building construction and be experienced in investigating and appraising subsidence and any resultant damage. An Expert should be capable of appraising and accepting responsibility for advising on some, or all of, the following:

- Severity and relevance of damage to buildings
- Scope of investigation required
- Design and/or specification of remedial works, including site inspections
- Building Regulations approval where necessary
- The recommendation to appoint an arboriculturist to advise on trees, tree physiology and water use (if required)
- Building design and construction
- Foundation design
- Ground engineering
- Written assurance on completion of remedial work, that the work has been properly undertaken in the form of a final report to the Insured and/or a Certificate of Structural Adequacy. The assurance should not rely solely on a guarantee or warranty issued by a contractor
- Overall project management

Experts will be central to establishing the cause of the damage or structural movement, by undertaking detailed technical investigation involving examination and measurement of the building and the damage, as well as instigation of any required geotechnical examinations of the ground on which the property stands. They should be capable of interpreting the results of those investigations and thus be able to distinguish between the effects and symptoms of subsidence and other possible causes of structural damage or movement.

The Expert can be directly appointed by an Insurer or indirectly via a Loss Adjuster, but most commonly are part of a Loss Adjuster's in-house technical team. Most Insurers will undertake responsibility for appointing Experts and rarely ask their Insured to do this. It is nevertheless very important that before any Expert is appointed by an Insured and investigations started, the Insurer is informed of the damage and that approval for any fees to be incurred by the Expert, has been given.

The types of professional advisers who may be able to act either as Experts or in support of Experts are:

Chartered (CEng) and Incorporated (IEng) Engineers
Members of the Institution of Structural Engineers (IStructE) or the Institution of Civil Engineers (ICE) have the most appropriate qualifications to conduct the initial appraisal and any subsequent further detailed investigation. The investigation requires a much greater technical expertise which is normally only possessed by appropriately experienced Chartered and Incorporated Engineers, who may be employed by a firm of consultants or Loss Adjusters.

Chartered Surveyors and Builders
As with Engineers, there are many disciplines within the surveying profession and, again, not all will have the appropriate skills, knowledge and experience to carry out an initial appraisal. For instance, Chartered Building Surveyors qualified within the Royal Institution of Chartered Surveyors are likely to be more suitable than Valuation, Fine Art or Quantity Surveyors of the same Institution, although all may appear to have the same level of qualification.

As an alternative to using Engineers, Chartered Building Surveyors and Chartered Builders with relevant expertise and experience may be used to specify and supervise above-ground repair works. Chartered Building Surveyors also have a role to play in relation to party wall legislation (Appendix C).

Loss Adjusters
These are professional advisers appointed by insurance companies to advise impartially on the relevance of the claim within the terms of the policy. They are required to:

- Investigate, quantify and administer the claim as quickly and efficiently as possible in accordance with the terms of the policy
- Present the views of the Insured impartially to the Insurer, even if they are not in agreement with those views

The appointed Loss Adjuster is usually a Chartered Engineer or Chartered Building Surveyor. If not, they are unlikely to handle the technical aspects of a subsidence claim and will advise on insurance matters only, referring all technical matters to in-house or external Engineers and Surveyors.

Arboriculturists
In cases involving assessment of the effects of trees on buildings or requiring remedial action to trees, it may be necessary to consult a tree specialist. These specialists should be qualified arboriculturists who are normally members of a relevant professional association.

Loss Assessors
Loss Assessors are not normally associated with subsidence claims because of the technical issues involved but, where engaged by the Insured to assist in the making of a claim, their fees would be met solely by the Insured.

It should be emphasised that a lack of formal qualifications may not necessarily prevent an individual carrying out the initial appraisal if they have the relevant experience, but it is imperative that any appointed Expert can identify and appraise subsidence and heave damage, and distinguish it from other types of building/structural movement or damage.

2.2 Mortgage process: the relationship between Lenders, Borrowers, Insurers and Valuers

2.2.1 The mortgage

A Mortgage Deed is an agreement between a Lender and a Borrower usually in connection with a loan for a particular property purchase.

The Lender agrees to provide a sum of money to the Borrower to be used in the new purchase/remortgage of a property in return for a commitment by the Borrower to repay the loan, with interest, over the agreed mortgage term. There will be other agreed terms and conditions including methods of repayment, the necessity to insure the buildings, the need to keep the property in good repair, a requirement to notify the Lender of any works which might affect the value of the property and the procedure if the Borrower defaults on repayments. All these items will be contained in the Mortgage Deed. Where a mortgage is given, the Lender has an interest in the property because it is the collateral or security for the loan, but has no right of interference with the purchasers' rights as Owner unless there is a failure to comply with the terms of the Mortgage Deed.

The existence of the mortgage is entered on the Charges Register at the Land Registry against the address of the property. The Borrower (Owner) cannot therefore dispose of the property by selling it without the prospective purchaser being aware that there is a legal charge over the property, and hence the Borrower does not have an unencumbered right to sell it.

During the course of the sale of a property, the vendor is normally required to complete a standard Law Society Preliminary Enquiries form about the property for the benefit of the purchaser. This form requires the vendor to state (among other matters) whether the property has, to the best of their knowledge, suffered from subsidence, heave or landslip.

2.2.2 The role of the Valuer

Lenders will usually obtain a valuation on a property being considered for a mortgage loan. A Valuer reporting to a Lender will always be professionally qualified, usually being a member of the Royal Institution of Chartered Surveyors. The Valuer will be expected to have knowledge and experience of the local geographical area in which valuations are carried out, and to work within the Practice/Guidance Notes for Valuers produced by RICS, subject to any specific requirements of the particular Lender. Typically the Valuer will:

- Inspect the property, including factors such as its location and condition, which could have a significant effect on the market value
- Advise the Lender of the open market value. This will involve obtaining as much comparable sales evidence as is available regarding transactions of similar properties in the period immediately preceding the inspection
- Provide an opinion on the estimated current reinstatement cost of the building for insurance purposes. This will include any garages and outbuildings with the cost of the site clearance and professional fees. In determining this sum, the Valuer will normally have regard to the figures produced annually by the BCIS (Section 2.1.2)

In all these obligations, the Valuer's duty is to the Lender. However, where the Valuer knows that the prospective purchaser is relying on the report, the Valuer may also have a duty of care to the prospective purchaser. This duty of care will relate to the same services the Valuer was instructed to provide to the Lender, and only to those services. This may include the reporting of major factors which are, or ought to be, obvious in the course of a reasonably professional visual inspection where these are likely to materially affect the value of the property.

Only Valuers are expert in marketability and value. If doubt exists, or if there are circumstances where the value of the property is likely to be reduced, the advice of a Valuer should be sought.

2.2.3 Insurance arrangements

A Lender providing a mortgage loan will require the property taken as security to be insured, and for the perils covered to include subsidence, heave and landslip. Insurers encourage Lenders to take a role in evaluating structural movement insurance risks on their behalf. As a consequence, many Lenders have arrangements with Insurers whereby properties can be automatically taken on risk by the Insurer without the need for individual cases to be proposed to them unless 'specified circumstances' exist.

Examples of specified circumstances could include where a property is affected by movement which appears to be continuing, or where there are circumstances which represent a risk of movement in the future, such as trees on shrinkable soils which are close to or within influencing distance of the property. In these cases, the Insurer may wish to consider the details prior to deciding whether or not to provide cover for the property.

In implementing these arrangements, Lenders will usually require their Valuers' reports to contain clear statements (often using 'standard phrases') as to whether any identified movement is considered to be continuing or long-standing/non-progressive, or as to future risk if this is thought to exist.

Borrowers who are arranging insurance themselves rather than through the Lender, and who have been supplied with a copy of the Valuer's report, are likely to use the Valuer's comments when completing the Insurance Proposal Forms. Alternatively, Borrowers would be advised to obtain their own report if continuing movement had been identified.

While structural movement requires special care both in assessment and resulting recommendations, it is important to remember that a Valuer is expected to make a judgment based only on a limited inspection. If it is obvious that movement is historic, has ceased and is unlikely to reoccur, the Valuer will report accordingly (Chapter 6).

However, in the event of structural movement which appears to be progressive, or if there is doubt as to the current or future position, the Valuer is likely to call for a report (Section 2.2.4) from a Structural Engineer, or possibly a Chartered Building Surveyor. In serious cases the Valuer may recommend the property be declined by the Lender.

2.2.4 Standard home purchase reports

Mortgage valuation report

This is a report produced for a Lender. In reaching conclusions on the value of the property, the Valuer should not have regard to the amount of the loan. The report should comment on the apparent condition of the property within the scope of the limited inspection required to determine value. Normally it will give an open market valuation, an opinion on the estimate of the reinstatement cost in the event of total destruction of the property and details of anything likely to cause damage to the property such as nearby trees, a steeply sloping site or a nearby watercourse. A comment on the future saleability of the property may be included.

Any obvious major defects observed which, if not dealt with promptly, may cause the condition or value of the property to decline, will normally be reported. Depending on its lending policy, the Lender may:

- Require remedial work to be carried out before mortgage monies are released
- Withhold a portion of the loan until work is completed
- Require an undertaking by the Borrower to carry out the work within a reasonable period

If it is unclear how extensive the works may be, or if the full extent of disrepair cannot be readily assessed, the Valuer may call for further investigation by an appropriate specialist and may recommend that no loan be made until the issues have been clarified.

The purpose of a Mortgage Valuation Report is to provide a valuation at a particular point in time. It involves a limited inspection only — of as much of the interior and exterior of the property as is accessible to the Valuer without undue difficulty. It has never been intended to be a detailed report on the condition of the property. The inspection will not cover sub-floors or the roof void, other than what is visible from the access hatch. There is no obligation to lift floor coverings, or to move furnishings or effects, or to test services. While value is always influenced by condition, the valuation can only have regard to the matters revealed by the limited inspection.

The intending Borrower may reimburse the Lender for its costs in obtaining the Mortgage Valuation Report but practice varies and often Lenders offer free valuations. The Lender will use the report to decide whether or not to lend monies based on the security of the property. It would be unusual however, for the Lender to make a copy of the report available to the prospective Borrower.

Borrowers should be encouraged to obtain their own independent and more detailed RICS Home Survey Standard-compliant reports[10], as these will provide a much clearer picture of the overall condition of a property than the basic valuation report, which is not a survey.

It is often possible for the Borrower to arrange for one of these reports to be prepared by the same Valuer/Surveyor used by the Lender, which may help to reduce overall costs. The following are various types of residential survey available to the Borrower:

- **RICS Home Surveys**
  There are numerous different forms of residential survey developed by the Royal Institution of Chartered Surveyors as follows:

  - RICS Home Survey Level 1
  - RICS Home Survey Level 2 (survey only)
  - RICS Home Survey Level 2 (survey and valuation)
  - RICS Home Survey Level 3

  These reports within the home survey suite are intended to provide general information on the condition of the property, with the focus on matters considered urgent or significant and those on which the client should take action before making a legal commitment to buy. Survey reports can be produced by using the standard RICS-approved templates or completed on a bespoke basis, subject to compliance with the Home Survey Standard.

  Home Surveys can be arranged directly through independent firms of Surveyors, or through some Lending institutions if a mortgage is being sought at the same time. In the latter case, the Surveyor will usually produce two reports for two distinct clients — a Mortgage Valuation for the Lender and, under a separate private contract for which terms of engagement will need to be agreed, a Home Survey Level 1, 2 or 3 report for the potential Borrower.

  The Level 1 survey is a cost-effective option, providing a potential buyer with general information on the condition of a building but which does not go into the sort of detail normally expected from a 'Building Survey'. It is, however, based on a more detailed inspection than that required for a Mortgage Valuation.

  A Level 2 survey is most suitable for conventional properties which are in reasonable condition. The report provides more detailed information than a Level 1 survey and offers the choice of either a survey or a survey and valuation. Opening up of the fabric of the building will not usually be undertaken, and the testing of services and other detailed investigations are also excluded. Some properties are too old and/or too large to be suitable for this kind of report, and in these cases a Surveyor may be unwilling to provide one.

  A Level 3 survey (previously known as a 'Building Survey' or 'Full Structural Survey') is likely to be well suited to large or older properties, properties in a poor state of repair, or properties which may be unusual or altered, or if major works are planned by the purchaser. This most comprehensive RICS Home Survey report provides an in-depth analysis of the property's condition and includes advice on defects, repairs and maintenance options.

  For all RICS Home Surveys there will be an opportunity for the client to discuss their requirements and concerns with the Surveyor before and after the inspection takes place.

- **Structural inspection report**
  This type of report may be recommended by a Valuer/Surveyor if there is evidence of structural problems, such as cracking, distortions or other signs of movement indicative of structural inadequacy or instability, which require more detailed investigation. It will normally be carried out by a Chartered or Incorporated Engineer, or a Chartered Building Surveyor.

- **Arboricultural report**
  This type of report may be recommended by a Valuer/Surveyor where vegetation is present which might cause or contribute to damage, particularly if the subsoil is clay (Chapter 8). However, it is most important that any Arboriculturist works closely with a Structural Engineer when assessing the influence of the vegetation on the structural stability of the property and any preventative or remedial work which needs to be carried out.

Terms of reference

It is very important that the purpose of the report is clearly set out as this will nearly always dictate the scope of the inspection and the cost of the report. Mortgage Valuation and Home Survey reports are usually subject to standard terms.

3 Stakeholder considerations

3.1 Introduction

Significant subsidence damage is normally insured under domestic insurance policies in the UK, along with other risks which can cause major damage e.g., flooding, storm, fire and explosion. If the property is less than 10 years old and has a latent defect, it may be insured on this policy as well. Chapter 4 discusses property insurance, warranties and guarantees in more detail.

Matters relating to subsidence and heave damage can be complicated, because many parties are involved between the discovery of the damage and the completion of the remedial works. In dealing with these damage claims and remedial works contracts, it is clear that different stakeholders have different concerns, and frequently view the problem differently.

This chapter identifies the stakeholders, other than the Expert, and highlights the interests and concerns of each. It is hoped that everyone involved will then appreciate the rationale behind the viewpoints and decisions of others, bearing in mind that the overriding aim should be that remedial work stabilises and repairs the affected property in such a way that insurance cover can be continued under normal terms i.e., the property should be restored to the same state it was in prior to subsidence or heave occurring.

Following remedial works, no 'blight' (perceived loss in property value, Section 4.6) should attach to the property, although in the unlikely event it does, note that any such perceived loss in value will not be taken into account in any insurance settlement because the wording of most policies refers to 'loss or damage'.

3.2 Customer care

Subsidence claims can be concerning to property Owners (the Insured), and Experts and contractors should be aware that their actions can significantly affect the relationship between the Insurer and the Insured. The following should be considered:

- Communication — good communication is vital, and time should be allocated for understanding an Insured's concerns, answering their questions, explaining how their claim will be handled at each stage and outlining the eventual scope of the works. Cordial relations should always be maintained between parties, including neighbours, who may have an interest, e.g., they may own trees that are implicated
- Confidence — establish the Insured's confidence in the fairness and competence of the Expert and other people dealing with their claim
- Respect — treat the Insured and their home with respect throughout the claim, particularly during the repair contract
- Quality — ensure that the contractor's labour and materials are of the required quality to meet the agreed specification
- Time — provide a proactive service to the Insured to ensure that as far as possible each stage of the claim is completed within agreed timescales
- Regulatory — Insurers are regulated and are expected to follow the FCA's ICOBS guidance[11]. This includes:
  - Handling claims promptly and fairly
  - Providing reasonable guidance to help a Policyholder make a claim, and appropriate information on its progress
  - Not unreasonably rejecting a claim (including by terminating or avoiding a policy)
  - Settling claims promptly once settlement terms are agreed

In addition, the Equality Act 2010[12] applies to all service providers in Great Britain. Where needed, service providers are required to make changes to improve service for those with disabilities. These changes or 'reasonable adjustments' are required whenever customers with a disability would otherwise be at a substantial disadvantage. The FCA have issued guidelines for the fair treatment of all vulnerable customers by financial (including insurance) firms[13].

3.3 Property Owner (the Insured)

The Owner may be concerned about:

- The cause of the problem and its physical implications, e.g., safety of the property
- Whether the damage is recent or long-standing
- The cosmetic appearance of the property (as a result of the damage, and possibly even after the damage has been repaired)
- The difficulties likely to be encountered during the repair, e.g., access, noise, dust etc.
- The duration of the stabilisation process
- The financial costs (and ascertaining who is responsible for bearing these)
- Whether any intermediate repair work will be undertaken during the investigation, e.g., easing doors
- Payment of the insurance excess (generally £1,000)
- How to appoint an Expert — although Insurers generally appoint and pay for a suitable Expert to act jointly and impartially on behalf of the property Owner/Insured and the Insurers
- The appointment of a suitable contractor to carry out remedial work
- Guarantees or warranties on the repairs
- Any future insurance cover
- Difficulties in selling the property and a reduction in its value
- Trees affecting the property being subject to Tree Preservation Orders (TPOs)
- The implications of their property or the adjoining property being a listed building or located in a conservation area, which may involve abnormal costs and statutory permissions
- The proximity of neighbouring properties
- Their own trees having influence on neighbouring properties and *vice-versa*
- The reasons for monitoring and the length of the monitoring period

3.4 Lender

In general, Lenders become involved with subsidence, heave or landslip damage, if and when notified by an Owner about such damage or the need to make a claim. When asking an Insurer for advice or asking for updates on the progress of a claim, they:

- Tend to be concerned about the effect of the problem on the value of the property, e.g., can the property be sold in its damaged state for at least the amount of the outstanding mortgage?
- May be concerned about the outcome of the claim affecting the relationship between the Lender and the Borrower
- May have concerns over who pays (especially in cases where the total cost for the remedial works may be more than the claims settlement amount agreed)
- May be concerned about whether the damage is actually insured at all, i.e., is the damage due to an insured risk? If it is not insured, the Owner may not be able to pay for the necessary remedial works, possibly affecting the value of the property and, in turn, the security of the mortgage
- May be asked to provide a mortgage on a property where a claim has been notified, and monitoring is being carried out to establish that there is a valid claim and no repairs have been carried out. In these circumstances Lenders will be reluctant to provide a mortgage since there is doubt whether a valid claim exists and that funds will be available for repairs to be carried out

3.5 Insurer

The Insurer may be concerned about:

- Whether the damage meets a certain minimum level and whether the damage is unlikely to get worse (i.e., is it worth the time and expense of a detailed investigation by an Expert?)
- Whether the damage has actually been caused by an insured risk — in this instance has the damage been caused by subsidence or heave of the site?
- The adequacy of the sum insured, and as a result payment of the correct premium
- Whether the ground movement and/or the damage occurred during the policy period or existed before cover was arranged

- The potential cost of appraisal, investigations and the eventual remedial work needed
- The method of how the excess will be paid
- Failure to comply with relevant policy conditions, such as late notification of the claim
- Possible recoveries against third parties, e.g., tree Owners and their liability insurers
- The operation of the Tree Root Claims Agreement (Section 4.4.2)
- Legislation e.g., Party Wall etc. Act[14], Construction (Design and Management) Regulations[15], Building Regulations[16–18], listed building/planning (Appendices C and E)
- Whether a structural Warranty Provider (or other insurers) are involved

3.6 Regulatory authority/utilities

The Local Authority or the relevant utility company may become involved if the health and safety of any person is threatened. This is likely to be the case if:

- The property is in a dangerous condition
- Service supplies to the property (e.g., a gas supply pipe or a drain) are damaged
- The ground beneath the property or in the immediate vicinity has suffered landslip, or is likely to do so in the future
- Any trees involved are subject to TPOs
- The property is in a conservation area, is a listed building, subject to a building preservation notice or is a scheduled ancient monument

Approval under the Building Regulations is required for underpinning, superstructure work and drainage installation. Minor repairs do not normally require approval, although a building control authority/approver may be consulted if there is any doubt.

The Local Authority could be involved if the building is considered to be dangerous or neglected — by serving statutory notices or, in extreme cases, by ordering emergency works to remove immediate danger.

Similar powers are available to the Local Authorities under the Housing Act 1985[19] (where the property is tenanted) or otherwise through the Housing Act 2004[20].

Local Authorities are responsible for administering TPOs and may need to be involved in tree removal or pruning. Trees located on the highway invariably belong to highway authorities, and if such trees are causing damage or posing a threat to property the authority will become involved. Local Authorities normally employ their own arboricultural staff to advise on such matters.

If a property is listed or in a conservation area, the proposed works may need listed building consent, with involvement of a Local Authority conservation officer and possibly Historic England, CADW (Wales), Historic Environment Scotland or Historic Environment Division (Northern Ireland). It is recommended that the advice of the Local Authority conservation officer be sought at the earliest stage. Breaches of this legislation could result in fines or even custodial sentences.

3.7 Neighbours

Residents of adjacent/adjoining properties may become involved due to:

- The proximity of the affected property to their property
- The possibility of the problem affecting their property either immediately or in the future if it is not dealt with appropriately and effectively
- Their property (the adjacent/adjoining property) being affected during the remedial works or subsequently as a result of the remedial works, e.g., a property may be underpinned and rendered stable, yet it is suspected that the adjacent/adjoining property may be founded on poor ground and so be likely to suffer damage in the future because of differential movement between the stabilised property and the unstabilised property
- The application of the Party Wall, etc. Act if an adjoining property is subject to underpinning and repair close to, or actually under a party wall
- Trees belonging to the affected property being responsible for damage/potential damage to the adjacent/adjoining property

- Trees belonging to the adjacent/adjoining property being responsible for the damage to the affected property
- The affected property having an effect on the stability and/or sellability of the adjacent/adjoining property

Appendix C covers these issues in further detail.

3.8 Contractor

The Contractor is contracted to the Owner to carry out the remedial works. These may range from cosmetic repairs, such as filling cracks and redecoration, to preventative measures such as removing or managing trees, or repairs to the foundations such as underpinning.

The works should be properly designed and administered by the Expert so the Contractor can undertake and manage the work efficiently. The Contractor will need to establish whether:

- They are expected to be responsible for the design of any of the remedial works
- The Expert who has designed or specified the works will be inspecting and administering the work, and whether they are available at short notice to advise on any difficulties encountered on site
- The site has been properly investigated and the cause of the problem established
- The extent and standard of the work has been properly specified
- They are responsible for access arrangements, materials storage and working hours
- Payments will be made on time in accordance with the contract with provision for paying the excess
- Any particular legislation applies[14–21]
- The responsibility for meeting the cost of the works is to be met either directly or indirectly by more than one party
- Payment for subsidence related work forming the contract will be paid for by the Insurer. Any other work privately agreed between the Insured and the Contractor will be paid for directly by the Insured, to the contractor. Such work may involve a separate contract
- The Contractor is likely to have to deal with any decoration or other work, outside the scope of the main remedial works contract
- A direct payment mandate will be signed

4 Property insurance, warranties and guarantees

4.1 Introduction

A significant majority of domestic subsidence damage investigations and repairs are funded by Insurers and Warranty Providers. While insurance and warranties/guarantees provide protection and reassurance to Homeowners, and give financial security to Lenders, they are principally contractual. However, like all contracts, there are terms and conditions. They specify a number of obligations of those involved, give direction as to how a claim may be progressed, and provide the Insurer or Warranty Provider with subsequent legal entitlements to recover their eventual outlays.

This chapter provides a high-level explanation of the key issues. To understand the position on warranties and guarantees it is first essential to note the differences. Although a guarantee may be 'insurance backed', it differs legally from an insurance policy. A warranty for this purpose is in effect a 'latent defects' insurance policy. Domestic and commercial insurances and warranty policies are regulated in the same way and in many areas there are similarities in process and application, e.g., the need to notify damage promptly.

From a practitioner point of view, the most likely criteria for whether a domestic property claim falls under an insurance policy or a warranty, is that of the timing of the damage relative to the age of the property.

This chapter also considers some of the practical issues around establishing a claim, the general process which follows, and matters such as dealing with historic damage and policy exclusions.

The explanations given are for general guidance only and do not necessarily reflect the views of all Insurers, Warranty Providers or Mortgage Lenders.

The majority of domestic property Owners have their properties insured against subsidence, heave and landslip. Owners of homes which are less than 10 years old are also likely to have a building warranty. Between them, insurance policies and warranties provide Homeowners with the peace of mind that anything unexpected affecting the property, which is covered by the building insurance or warranty, will be repaired, subject to agreed terms and conditions.

All new homes in England built for sale are required to have a minimum 10-year insurance backed warranty. Note, that the Building Safety Act[22] proposes to extend this to 15 years, but at the time of writing, secondary legislation is required before this can come into effect.

Lenders will have various lending arrangements such as 'prime' mortgages, 'buy to let', 'later life' and other equity products. Generally, they require properties to be covered by building insurances and warranties to protect their security. The general basis on which the purchase money is lent is that the property will be maintained in a readily saleable condition so that if it needs to be sold, there is the best chance a sum equal to or exceeding the amount of the loan will be realised and repaid.

It is in the interests of both the Homeowner and the Warranty Provider that the quality of the house is satisfactory and defect free. This is why Warranty Providers place such emphasis on the quality of construction. Where there is a defect or structural damage that is covered by the warranty within that period, it is the Warranty Provider that principally ensures that the defect or damage is rectified.

Not all damage is covered by domestic property building insurance, e.g., damage caused not by foundation movement, but which has solely arisen from:

- Poor quality building work
- Inadequate materials
- Inadequate design
- Failure to maintain the property reasonably

Examples may include lintel failure, roof spread, corroding steelwork or thermally inefficient windows. This kind of damage is often expensive to repair and the lack of insurance cover may not be readily understood by Owners. Insurance backed warranties may respond to deal with such problems on new properties, but in older properties such matters are maintenance issues which are not covered by a general insurance policy. For Lenders, these matters are mainly only of interest in a default or repossession situation, although often the situation can be complex.

The key point is that some items may be uninsured or fall outside the scope of a warranty, and for this reason it is important that there is an adequate structural survey at the time of purchase.

If a Homeowner is concerned that cracking may be subsidence related it would be prudent for them to notify both their Insurer and Warranty Provider so that the matter can be investigated and confirmed promptly.

4.2 Insurance and warranty contracts

This section explains the basis of both insurance and warranty contracts. Many of the insurance principles discussed in the following subsections also apply to warranty covers, as they are insurance backed products. Insurance policies and warranty cover provide Homeowner benefits, but are different in terms of legal meaning, specific wording, extent of coverage, and process, even if the warranty itself is insurance backed. For this reason, Insurers and Warranty Providers do not normally 'share' the cost of any repair, which is a legal process known as 'contribution'.

4.2.1 The insurance contract

An insurance policy is a legal contract between two parties, namely:

- The insurance company, known as the 'Insurer'
- The Policyholder and, where applicable, the Lender, known as the 'Insured' as both usually have an insurable interest in the insurance policy

Under a typical domestic insurance policy, a property Owner insures their property against specified events or 'perils', such as subsidence, heave and landslip. Commercial property insurance cover is often significantly different from domestic property insurance in terms of scope, conditions and exclusions.

4.2.2 Purchasing insurance cover

Whereas warranty cover might normally be bundled into a purchase arrangement, the buying of building insurance allows a degree of choice on the part of the Homeowner in terms of the provider and scope of cover.

Buildings insurance covers the cost of repairing the damage to the structure of a home as well as any fixtures and fittings such as bathrooms and kitchens. It is often sold alongside contents insurance as part of a wider home insurance policy. Buildings cover provides for a wide range of risks, and overall will usually include the cost of repair, demolition, site clearance, professional fees, and planning and building regulation charges.

Additional cover will be available for 'alternative accommodation' should the home become uninhabitable. The additional costs of complying with Building Regulations to the damaged part of the property may also be covered.

In the UK, the marketing and selling of buildings insurance are regulated by the Financial Conduct Authority (FCA).

4.2.3 Basis of insurance claims settlement

An insurance policy protects the Policyholder against loss or damage caused by one of the insured events. In simple terms, the Insurer will return the Insured to their position prior to the insured event (a process known as 'indemnity', although in recent times this has been extended to disregard the age and condition of the building, subject to certain terms and conditions).

This can be done in one of two ways:

1. Paying for the cost of repairing the damage.
2. If it is not economically viable to repair the damage, a settlement equal to the diminution in market value may be paid, i.e., the Insured may be paid the difference between the market value of the property as if it had not suffered the damage and the reduced market value of the property taking the damage into account.

Many policies have additional contractual provisos, including that settlement is based on the repair cost (which is the lower of either the cost of the work if the Insurer's nominated contractor carried out the work, or the cost of the work as competitively estimated or tendered by the Policyholder's builder). In addition, clauses may refer to VAT being withheld in a cash settlement unless evidenced by VAT invoices.

Some commercial property policies may also have specific clauses referring to the Insurer's right to directly carry out the reinstatement within stated provisos.

There may also be restrictions on cover if the buildings have not been well maintained, or if the sum insured does not reflect the full reinstatement cost.

While repair and redecoration may give rise to improvements, it is not the intention of the policy to pay for betterment (Section 4.4.5) or maintenance of the fabric of the building above and beyond that necessary to carry out reasonable repairs and redecoration following subsidence, heave and landslip movement.

It is important to remember that the extent of policy cover and basis of financial settlement will depend on the individual policy wording.

4.2.4 The warranty contract
The warranty is taken out by the builder or developer to protect the homebuyer against problems that might occur with their home both before and after completion of construction.

There are 3 general areas of warranty or guarantee:

1. Contractor guarantee — backed by the contractor with no insurance behind it (other than possibly PI insurance).
2. Contractor Insurance Backed Guarantee — backed by an insurance policy that will benefit the client if the contractor ceases to trade for any reason, but there is still a need to prove negligence, which if contested can be risky and expensive.
3. Warranty (latent defects insurance) — generally for a 10-year term with the first two-year's of liability residing with the contractor (if the contractor ceases to trade in the initial term, the policy would react). The policy normally attaches to the property so there is no need to transfer, nor a need to prove negligence in order to make a claim; simply a need to demonstrate that the claim exists e.g., that a crack exists and that it is related to structural movement rather than thermal movement.

Examples of what is typically covered include:

- The developer becoming insolvent before completion — in which case the buyer's deposit is protected
- Defects in the construction, such as drains not connected or the heating not working, that come to light in the first two years after completion
- Major problems with the structure of the home including foundations, walls, roofs, ceilings, chimneys and load-bearing parts of the floors, that come to light 3–10 years after completion

It is important for buyers of new homes to check the provider of their warranty, when the warranty cover starts, and what is actually covered, as the detail of coverage varies between providers.

The warranty contract will ultimately be between the buyer of the new home and the Warranty Provider. Most warranty polices are transferable between Homeowners if it is sold within the term of the policy. The contract is usually drawn up and exchanged as part of the sale of the home during conveyancing, and it is important for Homeowners to be aware of their Warranty Provider and cover and keep their policy document safe.

There are currently many providers of insurance backed warranties. In the UK, the National House Building Council (NHBC), LABC Warranty and Premier Guarantee are the leading providers.

4.2.5 Basis of warranty claims settlement

Where a defect is covered by the terms of the warranty, the Warranty Provider will normally arrange for the original builder to return and rectify the problem if it is within two years of construction. Where the damage occurred 3–10 years after construction, the Warranty Provider will pay to put right the defect.

4.2.6 Handling of complaints

All Insurers and Warranty Providers have a documented process in the event of disagreement or customer complaint.

The management of insurance and warranty claims is a regulated activity under the Financial Services and Markets Act 2000 (FSMA)[23]. This legislation requires Insurers to provide 'reasonable guidance' to help a Policyholder make a claim.

The handling of complaints arising from the claims handling process is governed by the FCA, defining how complaints are recorded and responded to. Customers have access to the Financial Ombudsman Service (FOS) in cases where they are unhappy with how their Insurer or Warranty Provider is handling their complaint[5].

Table 4.1 highlights the principal similarities and differences between insurance cover, warranty cover and guarantees.

Table 4.1: Differences between general property insurance cover, warranty (latent defects) cover and guarantees

Aspect of cover	Insurance	Warranty (latent defects)	Guarantee
Scope	Provides compensation ('indemnity') against specific events (perils) occuring e.g., subsidence. Not all insurance policies refer to the term 'subsidence'. Some insurance policies cover 'settlement to a building which is more than 10 years old'.	Provides compensation for building defects even if these are unrelated to the operation of an insured event, i.e., subsidence. Warranties do not deal with 'all defects', but are specific regarding what is not covered e.g., cosmetic damage where the structural integrity of the building is not affected.	Can be insurance backed, and is based on the contract between client and developer/contractor. There is a need to prove negligence in order to pursue a claim, which can involve costly investigations and reports — the cost of which cannot always be recovered from the liable party.
Timing	No timeframe but Insurers expect Warranty Providers to deal with issues within 10 years of construction unless there are exceptional circumstances or significant and unforeseen external events or situations.	Looks to deal with issues within 10 years of original construction.	Reliant on contract law — six years for defects in a contract under hand or 12 years under seal — the guarantee may extend these terms.
Limitations	Will not provide cover in respect of defective design or workmanship, which is usually specifically excluded by the Insurer.	Will provide cover for defective design and workmanship, subject to agreed terms and conditions.	Necessity to prove negligence, which is time-consuming and costly.

(continued)

Table 4.1: *continued*

Aspect of cover	Insurance	Warranty (latent defects)	Guarantee
Investigation process	Where an Insured suspects subsidence, in most cases an Insurer will instruct and pay for one or more Experts to investigate. This is mainly for Insurers to establish whether damage is due to the operation of an insured peril, i.e., 'subsidence', and where there is cover, to agree settlement of the claim within the terms of the policy. Additionally, the Expert will gather evidence to take action against a third party e.g., a negligent builder.	Where a Homeowner suspects a valid claim under the warranty, in most cases the Warranty Provider will instruct an Expert to investigate. The Warranty Provider will usually pay for the cost of this initial appraisal, although the criteria for the inspection may be different from that of an insurer, e.g., it is not important to identify the operation of an 'insured peril' such as subsidence.	Generally the onus is on the Homeowner to prove a valid negligence claim and try and to recover the costs retrospectively.
Professional representation by Homeowner	If an Insured wishes to retain the services of a specialist consultant as a separate adviser to give a second opinion, this should be acceptable to all parties, but the Insured may have to pay for this separately. The Insured may also decide to appoint a claims management company (CMC) to represent them, and insurance policies do not generally cover their fees. In the UK, the activities of CMCs are also regulated by the FCA.	If a Homeowner wishes to retain the services of a specialist consultant as a separate adviser to give a second opinion, this should be acceptable to all parties but it is likely that the Homeowner may have to pay for this separately.	The costs of representation are borne by the Homeowner with the risk that they cannot be recovered as part of the claim.
Transfer	An insurance policy per se cannot be transferred from one party to another, although the 'benefits' of any outstanding claim can be legally assigned to a third party e.g., a Homebuyer.	A warranty can be transferred from one party to another, usually as part of a homebuying process.	Guarantees are generally to the contracted parties and therefore have to be transferred on a sale. This often involves an administration fee.
Policy excess	Most insurance policies carry a policy excess or 'deductible', which requires the Policyholder to pay an amount specified under the policy.	Claims occurring within a minimum value as specified under the individual warranty schedule are specifically excluded.	Guarantees rarely carry a minimum claim value or excess.

(continued)

Table 4.1: *continued*

Aspect of cover	Insurance	Warranty (latent defects)	Guarantee
Tree and drainage management costs	Insurers are usually prepared to deal with the reasonable cost of tree or drainage management, provided the trees or drains are owned by the Insured and are on their land. Insurers may also consider funding the cost of management of a neighbour's trees or drains — however, the position would need to be confirmed on a case by case basis.	Warranty Providers are usually prepared to deal with the reasonable cost of tree or drain management, provided they are owned by the Insured and are on their land. They are normally unwilling to fund the cost of management of a neighbour's trees or drains.	N/A
Subrogation or 'Right of Recovery'	Under the terms of the policy, Insurers are entitled to 'stand in the shoes of the Policyholder' and recover their outlays against a negligent Third Party such as a builder or tree Owner.	Warranty Providers are usually entitled to a 'Right of Recovery' against a third party.	A claim under a guarantee could involve other parties to the negligence but full costs are often not recovered by the Homeowner.
Alternative accommodation	Insurers usually provide cover for the cost of alternative accommodation when the property becomes uninhabitable, incurred with prior agreement and subject to contractual limits.	Warranty Providers usually provide cover for the cost of alternative accommodation, incurred with prior agreement and subject to contractual limits.	Part of the costs of a claim, but would be subject to the contract terms prevailing.
Sharing of costs	In certain circumstances where there are multiple policies in place, Insurers will share the cost of repair or, more likely, the cost of alternative accommodation. This process is known as 'contribution'.	Warranty Providers do not normally share costs with Insurers or other Warranty Providers (as there is usually only one warranty operating at any given time).	No costs shared.

4.3 Cover for subsidence damage

This section explains the scope of insurance cover and the key conditions and limitations. Although insurance policies and warranties differ, there are similarities in terms of conditions and processes.

The insurance policy is a contract between two parties — the Insured and the Insurer, and is renewable on an annual basis. Similarly, a warranty is a contractual agreement between the Homeowner and the Warranty Provider. Parties are bound by the precise wording of the contract between them. They are entitled to rely on the wording of the contract and entitled also to the terms and conditions of that contract as appropriate.

The terminology/wording used by Insurers and Warranty Providers differs, and small differences in the policy/warranty can either expand or restrict the level of cover. It is important therefore that all parties fully understand the precise wording.

Some examples of typical exclusions seen in domestic household insurance policies are:

- Damage resulting from demolition, structural repairs, alterations, or maintenance to buildings
- Damage resulting from faulty design and/or workmanship or the use of defective materials in buildings

- Damage to groundbearing floor slabs, e.g., floor slabs supported by hardcore or other fill material, or from solid floor slabs moving, unless the foundations beneath the external walls of the building are damaged by the same cause and at the same time
- Damage to outdoor swimming pools, tennis courts, patios, terraces, hot tubs, ponds, drives, footpaths, walls, gates, hedges or fences unless the main property, an outbuilding or garage is damaged by the same cause and at the same time
- Damage caused by coastal or watercourse erosion
- Damage caused by structures 'bedding down' or settlement of newly made-up ground (sometimes qualified by being within 10 years of original construction)
- Any payment within the policy excess

There are also specific 'claims conditions' within the insurance policy, such as the need to notify Insurers or Warranty Providers of a claim as soon as practically and reasonably possible. A failure by the Insured to comply with these conditions may entitle the Insurer or Warranty Provider to refuse to deal with the claim. The effect that the breach of any of the conditions by the Insured might have on a claim is explained in the following subsections.

In a domestic situation, Insurers will seldom decline to deal with a claim due to breach of a policy condition if that breach has been innocent in nature and not material to the loss and does not prejudice the Insurers' position. They may, however, require a detailed explanation before a decision is reached. These may involve lengthy investigations. Occasionally the expression 'without prejudice' will be used, which means that Insurers are investigating a matter without committing themselves legally or contractually.

4.3.1 Insurance cover and continuity between Insurers

Subsidence insurance is intended to cover the Policyholder against damage to the insured buildings caused by subsidence or heave of the site 'on which the buildings stand', or landslip subject to the policy terms, definitions, conditions and exceptions (Chapter 2). Note that some insurance policies do not specifically refer to subsidence but rather to 'settlement'. In those cases, they often exclude settlement to buildings less than 10 years old. Some policies give clear definitions of subsidence, landslip, heave and settlement but others may not, so care needs to be taken. The FOS also provides some guidance on the differences between subsidence and settlement[5].

In principle, Insurers are only contracted to cover subsidence, heave or landslip damage which occurs after the start date of the cover. As a result, damage which originated before the cover start date may be paid partly or wholly by the previous Insurer, depending on how long the current Insurer has been on risk (i.e., how long their policy has been in operation).

Because of the potential for disputes over the timing of damage, the Association of British Insurers (ABI) developed the *Domestic subsidence/heave/landslip "change of insurer" claims agreement*[24] which sets out rules for handling claims notified within 12 months of policy inception. This guidance is subject to periodic review so reference should always be made to the ABI website to ensure the latest version is applied.

This is not normally an issue for Warranty Providers who have an interest from the date of construction.

4.3.2 Establishing a claim under an insurance or warranty policy

In establishing whether damage has arisen due to an insured peril such as subsidence, heave or landslip, consideration needs to be given to whether there has been some external factor, or combination of factors, which have influenced the ground beneath the foundations. Experts will therefore seek to focus on identifying what might be that 'cause' or 'external factor' such as the influence of trees or drains.

The expression 'proximate cause' is also used by insurers to describe the underlying/most significant cause of the damage.

It is not important at this initial stage to know the 'exact cause' of the movement to enable a decision to be made as to whether damage is the result of subsidence. If cracking to the property is wholly consistent with ground movement because of an external factor, then it may be possible to conclude *'on the balance of probabilities'* that movement of site must be giving rise to the damage which is occurring, and this may be sufficient to satisfy a domestic property Insurer that a claim is admissible, subject to all other conditions being complied with.

However, a more precise understanding of the cause of damage will need to be determined to allow appropriate remedial measures to be correctly designed (Chapter 7). Evidential standards will usually also need to be high, especially where there are matters of legal responsibilities or subrogation to consider, or where Insurers need to demonstrate an applied exclusion which may invalidate the claim (Section 2.1.2).

Establishing a claim under a warranty does not require the identification of a specific peril e.g., subsidence, but principally is a matter of understanding the cause of damage and whether it is included or excluded under the terms of the warranty document.

It must be emphasised, however, that it was never the intention of Insurers or Warranty Providers to pay to remedy damage which is an inevitable consequence of the inherent nature of a building, e.g., damage arising from regular thermal, minor seasonal or cyclical movement. This is as much a part of the normal and predictable characteristics of a building as, say, the swelling and shrinkage of woodwork which results in doors and windows sticking from time to time and then becoming free with a change of season.

4.3.3 Timing of subsidence damage

Difficulties can arise where subsidence-related movement — which initially seemed self-controllable and non-progressive at the time cover started — continues, and results in further damage. In such circumstances, it may be impossible to repair the more recent damage without also repairing damage caused by the previous problem, such as in the case of partial underpinning (Section 9.5). In essence, the issue is whether this is a new claim, or an existing (unresolved) claim.

Evidence of continuing subsidence, heave or landslip has a major impact on saleability and value which is of concern to both Lenders and Owners, particularly those who wish to buy or sell in the near future. It follows that investigation of subsidence, heave or landslip problems must be addressed quickly, that repairs must be satisfactory from an engineering point of view and that saleability should be restored.

Where saleability is likely to be permanently affected, Insurers and Warranty Providers may decide to make a once-and-for-all payment to cover diminution in value to fully restore the Insured to their previous position and discharge any financial obligations, rather than incur the cost of repairs. But in such cases and for obvious reasons, future cover for subsidence, heave or landslip would be excluded from the policy.

Normally, Insurers will accept that a claim will be considered under their policy, provided subsidence, heave or landslip is evident and structural damage has occurred during the currency of their policy or warranty, and that no exclusions or breaches of condition apply.

Insurers and Warranty Providers will not pay for preventative measures such as underpinning, where no damage to the relevant part of the property has arisen.

4.3.4 Reopening historic claims

Occasionally, further damage may appear in a property after repairs have been completed. Whether this new damage will be regarded as a new claim or a continuation of the earlier claim will be determined by the technical evidence. The principal test will be whether or not there has been a new ‘event’ or intervening factor. This can be a contentious issue.

It is the view of Insurers and Warranty Providers that any new damage will be considered as a new claim (for example, appropriate excess being payable) unless the new damage has appeared in the same part of the property which had previously been repaired and is attributable to the previous cause. Should a subsequent spell of dry weather occur for example, this may not be regarded as the same cause, bearing in mind as a comparison that in the case of storm damage a second storm would be regarded as a separate incident.

Generally, for a claim to be reopened, the new damage must also appear within a relatively short period of time (probably no longer than two years) although some Insurers may apply different time-based criteria (possibly five years). The FOS has indicated that repairs should be ‘effective and lasting’ and have suggest a 10-year period.

Reoccurrence of damage may also be associated with defective design or construction of the repairs carried out in the original claim. In that case the Insurer or Policyholder may seek a remedy from the original designer or contractor, depending on the contractual relationships that exist between the various parties, and subject to limitation periods which are generally six years from when the breach of contract occurred.

4.4 Conditions and exclusions

4.4.1 General

Although insurance policies and warranties differ in legal nature, there are similarities in terms of the conditions and exclusions which apply. Reference should be made to the specific wordings of the policy/warranty.

4.4.2 Prompt notification

When the Insured becomes aware of any damage, they should notify the Insurer or Warranty Provider as soon as possible and provide all the information that is reasonably required by the Insurer, so that:

- Steps can be taken to commence the investigation and perhaps also prevent the damage from getting worse
- Prevent delays that may lead to the cost of the work increasing because of inflation or deterioration

When dealing with a claim, the Insurer or Warranty Provider is entitled to take over whatever legal right the Policyholder might have to recover the costs resulting from damage, perhaps from a negligent builder or the Owner of a tree which has caused the damage. Such action is known as a 'subrogated' action (Section 2.1.2).

It is important that, for example, an involved tree Owner and their liability Insurers are placed on notice as quickly as possible, both for legal reasons and also to allow the tree Owner's Insurers the opportunity to deal with the matter without their own position being prejudiced by delay. The ABI's claims agreement[25] deals with this, but note it does not apply to statutory bodies or commercial entities, relating only to domestic properties which are owned/tenanted/occupied by an Insured in a personal capacity.

4.4.3 Non-disclosure

When a customer buys or renews an insurance policy or warranty in the UK, CIDRA regulations require the customer to take reasonable care not to make any misrepresentation[26].

Insurers have an obligation to ask clear questions at policy inception and renewal. Should the Insurer not be provided with accurate information, the Act lays out the solution, dependent on whether any misrepresentation was deemed to be deliberate, careless or innocent. For example, if a customer deliberately fails to disclose information, a criminal offence may have been committed under Section 3 of the Fraud Act 2006[27]. It is important for the customer to consider:

- What type of policy is being taken out
- What documents have been provided by the Insurer or Warranty Provider
- What specific questions have been asked by the Insurer or Warranty Provider
- Whether an agent (e.g., a broker) is involved

With regard to subsidence damage, the questions asked by an Insurer will likely relate to a specific time period (typically the previous 25 years) and could include:

- Has a claim been made for subsidence, landslip or heave?
- Are there any signs of damage resulting from subsidence, landslip or heave?
- Has your property been monitored as a result of subsidence, landslip or heave?
- Has your property suffered from structural movement or settlement as a result of subsidence, landslip or heave?

Depending on the response, an Insurer may ask additional questions regarding the particular subsidence incident referred to before deciding whether to offer cover. At purchase, a building survey may identify signs of settlement or movement but record it as only 'historic' or 'long-standing' and 'non-progressive' or 'acceptable'. If such comments also mention the possibility of subsidence, it may prompt more questions or underwriting referral.

In some cases where damage is apparent from a purchaser's valuation, or from a claim that has not yet been registered, arrangements may be made for the seller's Insurer to continue to Insure the property. Also, there are specialist Insurers who provide cover for properties which have suffered subsidence.

It should be noted that in the UK, only around 25% of purchasers with a mortgage seek a RICS Home Survey Level 2 or higher. Therefore, many problems at the point of sale are never captured. The increased use of automated valuations and desktop assessments means that in many cases Homebuyers are not aware of the whole picture, leading to issues where damage clearly predates purchase. Prudent house buyers arrange an appropriate and detailed survey (Section 2.2.4).

4.4.4 Maintenance and repair

A property Owner (the Insured) who has been made aware that a defect needs repairing, or that a particular course of action needs to be followed to minimise future risk, has a responsibility to mitigate the damage, that is, to avoid or lessen the occurrence of further damage by implementing the appropriate measures. This might include, for example, tree management or drainage repair.

Insurers and Warranty Providers will expect this to be done even if they neither required the relevant issue to be notified to them, nor made it a condition of insurance. In effect, it is a 'duty of care' to take reasonable steps to maintain the property.

If the Insured has failed to reasonably maintain the property, this could affect the outcome of the claim, as it may comprise a breach of one of the general conditions of the policy. Failure to reasonably mitigate such damage or its consequences may permit Insurers and Warranty Providers to reduce the amount they pay under the claim, or to avoid dealing with the claim altogether.

4.4.5 Betterment

Where the Insured has not reasonably maintained the part of the property which is the subject of the claim, or has sought an improvement in specification which is not required to comply with regulations, Insurers and Warranty Providers may require a contribution from the Insured towards 'betterment' or improvement.

4.4.6 Drives, footpaths and garden walls

Typically, most insurance policies exclude damage to drives, footpaths and garden walls unless the main dwelling is affected by the same cause and at the same time. However, policy wording referring to 'same cause and at the same time' may vary and should be checked carefully.

4.4.7 Floor slabs

Typically, most household insurance policies specifically exclude from subsidence cover:

> *"damage to groundbearing floor slabs, e.g., floor slabs supported by hardcore or other fill material, or from solid floor slabs moving, unless the foundations beneath the external walls of the building are damaged by the same cause and at the same time".*

If the slab supports loadbearing walls, cover may apply subject to the usual exclusions, and the precise policy wording should be confirmed in each case. This is not strictly true in the case of warranties.

Lenders and Valuers should appreciate that minor amounts of slab movement are rarely of any consequence from a structural and serviceability point of view.

4.4.8 Subrogation and other key insurance principles

Subrogation rights, contained within the insurance contract, enable the Insurer to recover all or part of the claim costs from the third party who caused or contributed to the damage.

In this context the Expert will need to ensure that the scope and detail of the investigation meets the likely evidential thresholds required to pursue a legal claim against potential third parties. The Insured is expected to support such an action, especially as it is being done 'in their name', which Insurers are entitled to do under the policy. The Warranty Provider usually also carries a similar provision.

Other key insurance principles are referenced throughout the chapter.

4.4.9 Adequacy of sum insured
The sum insured, which is the responsibility of the Insured, must reflect the cost of rebuilding a property. If it does not, Insurers may have grounds for adjusting their liability under the policy — such as taking age and condition of the property into account when agreeing settlement, or even applying a *pro rata* settlement relative to the degree of under-insurance.

Most household insurance in the UK is now factor-rated, providing a sum insured typically ranging from £400,000 to an unlimited amount, depending on the contract. This greatly reduces the need for Policyholders to make an accurate sum insured calculation.

Factors typically include:

- Postcode
- Number of bedrooms
- Number of bathrooms
- Age and type of property
- Construction type (referred to as 'standard' or 'non-standard' construction)

Customers in the UK can find out more about calculating the rebuilding cost of their home from the ABI's website[28] which is based on published rebuilding cost information published by the Building Cost Information Service (BCIS). It is important to recognise that this service provides a rebuilding cost guide based on minimal information rather than a detailed assessment, and may not take into account special features or unusual market conditions relative to labour or materials.

4.5 Handling of subsidence, heave and landslip claims

This section provides an outline of the processes involved in the investigation of property insurance and warranty claims. Although most Insurers and Warranty Providers have their own processes, there are common features that generally apply.

4.5.1 Investigation of property insurance claims
When a subsidence, heave or landslip claim is made by the Insured, it is essential that it is properly investigated. The following parties could be involved:

- The Insured — usually the property Owner (or their appointed advisor)
- The Insurer
- Any of the Experts listed in Section 2.1.3

Often insurance is in the joint names of the Lender and the Borrower, but Lenders rarely get involved in any investigation process. The interest of Lenders is principally in ensuring that their financial security or equity in the damaged property is maintained.

Experts have a duty to offer a professional opinion on the matters referred to them.

> The Insured should understand that in the context of subsidence, heave and landslip, engineering input is particularly appropriate to the problems which are occurring below ground, with surveying input being more appropriate to the problems which are evident above ground.

Both the engineering and surveying inputs could be made by one Expert or a multi-disciplined firm. Not all Engineers and Surveyors are suitably qualified and experienced to deal with subsidence, heave or landslip damage (Chapters 2, 6 and 7), and even fewer are competent to deal with the legal issues that may arise.

4.5.2 Investigation of warranty claims
Normally an investigation will be conducted by the Warranty Provider who will fund the cost of any tests. The use of a suitably qualified and experienced Expert will hopefully lead to a prompt and accurate diagnosis.

4.5.3 Investigation procedure

Where a Homeowner suspects subsidence, in most cases an Insurer or Warranty Provider will instruct an Expert to investigate. In this situation, the Insurer or Warranty Provider will usually pay for the cost of this initial appraisal, the purpose of which is explained in detail in Chapter 6.

The Expert will:

- Determine whether the damage arises from subsidence, heave or landslip, or whether damage is due to some other cause which may or may not be covered under the policy or warranty
- Be required to comment on whether there is a possible breach of any policy condition (which might require further investigation) and/or the impact of any policy exclusions
- Collect evidence should an action against a third party be needed

If an Insured wishes to retain the services of their own specialist consultant as a separate adviser to give a second opinion on any particular aspect, this should be acceptable to all parties. However, some Insurers may require that the Insured pays for this separately from their own funds. Insureds and Homeowners who are considering such a course of action should liaise with their Insurers or Warranty Provider before incurring any costs which may not be recoverable under a claim.

Based on the initial visit, further detailed investigation may be needed so that the precise cause of the problem can be identified. Bearing in mind that subsidence, heave or landslip often results from external factors, such as leaking drains, unmanaged or inappropriately sited trees, etc., efforts should be made to identify them and remedy any problems of the site so that the foundations can be stabilised. This will be particularly important if third parties are implicated through tree ownership. If, ultimately, the control of these factors does not work, or there is a lack of cooperation, consideration may have to be given to stabilising the foundations by underpinning, depending on the severity of the damage.

Investigations can be lengthy and often involve monitoring the property for 12 months or longer. If the Insured wishes to sell the property prior to completion of monitoring and repairs, it may be possible to transfer the insurance rights to a prospective purchaser. However, this will depend on the co-operation of several interested parties, as well as other factors such as the price being paid for the property, and whether liability has been accepted by the Insurer or Warranty Provider.

Documentation of the repair should include certification by the Expert at completion of the remedial work (Section 10.8 and Appendix B). It is strongly recommended that the Insured retains all documents relating to the repairs, as subsequent purchasers may well ask for details.

4.5.4 Claims handling and regulatory requirements

In the UK, the handling of claims is regulated by the FCA, and Insurers are expected to follow the rules laid down in the FCA's *Insurance Conduct of Business Sourcebook* (ICOBS)[11]. Claims handling rules include:

- Handling claims promptly and fairly
- Providing reasonable guidance to help a Policyholder make a claim, and appropriate information on its progress
- Not unreasonably rejecting a claim (including by terminating or avoiding a policy)
- Settling claims promptly once settlement terms are agreed

The Equality Act 2010[12] applies to all service providers in Great Britain, including Insurers, whether a charge is made for the service or not. Service providers are required to make changes, where needed, to improve service for disabled customers. These changes or 'reasonable adjustments' are required wherever disabled customers would otherwise be at a substantial disadvantage compared with non-disabled people.

In this context the FCA have set out guidelines for the fair treatment of all customers making an insurance claim.

4.5.5 Simple stabilisation measures — general principles

Detailed information on remedial measures is given in Chapter 10.

Technically, the cost of tree management or removal is not covered by the insurance policy and consequently would have to be carried out at the Homeowner's expense. However, most Insurers will pay for the reasonable cost of tree works as part of the stabilisation process. In the case of damage caused by trees, once the property has been stabilised and repairs carried out, it would then be the Owner's continued responsibility to manage the trees to prevent future damage.

Where leaking drains are a factor and the drains have not been accidentally damaged (e.g., damage due to defective workmanship or an external intervening cause) the insurance policy may not cover the drainage repair.

However, Insurers will generally pay for drainage repairs where it is important to help stabilise a property. 'Accidental damage to drains' i.e., damage due to a specific external force such as tree roots, is usually covered under a separate section of the policy and subject to a different policy excess. Whether subsidence damage has been caused by leaking drains is a matter of professional opinion and how Insurers respond may be a matter for their particular wording or interpretation.

Where the cost of continuing tree management or repair to drains is at the Homeowner's expense and there is a failure to carry out recommended preventative works, this could influence the Insurers' decision to deal with the claim and provide future cover. This approach is no more than a reinforcement of the duty of care by the Homeowner or Policyholder to protect their property, whether or not insurance is in place. Once drains have been repaired and trees pruned or removed, a period of monitoring will be required. The cost of the monitoring will normally be funded by the Insurer.

Parties should recognise that the longer it takes for the trees to be attended to, the worse the damage is likely to get, and the more expensive the repair is likely to be. It is very rare for Insurers to fund the cost of management of trees or repair of drains not owned by the Insured. There are often issues of legal ownership and other complex issues to consider.

On completion of investigation works, and provided that the property is stabilised by the control or elimination of any influencing factors, a schedule of repairs will normally be prepared, and contractors invited by the Expert to quote for the work. Repairs may be carried out by the Insurer's own contractor, or, with agreement, the customer's contractor. All this, subject to the deduction of any policy excess, will be undertaken at the Insurer's expense.

Where the repairs are carried out by the customer's contractor, or if reinstatement is being carried out in a different form, a financial settlement may be agreed. Policy conditions may treat the VAT aspect separately, withholding payment pending receipt of a VAT invoice. Where structural repairs are involved, Lenders may require evidence of the work being carried out but in many cases they are not usually engaged with the process, and rely on the Homeowner.

Where repairs are carried out, Insurers will generally continue normal cover, and in the majority of cases would be prepared to offer insurance cover to prospective purchasers (Section 4.6). Where a subsidence damaged property has been properly dealt with, and a Certificate of Structural Adequacy (Section 10.8 and Appendix B) is available from an appropriately qualified Expert, there should be no technical reason why any Insurer should refuse or remove cover.

4.6 Continuity of insurance and warranty cover after remedial works

This section deals with the issue of continued insurance cover after remedial works. Continuity of cover is rarely a problem for the warranty, as cover is not normally transferred between Warranty Providers unless one stops trading.

Cover on buildings which have experienced subsidence and have been underpinned, in whole or in part, is usually continued, provided the cover is to the same Owner. It should be appreciated that the Insurer may reasonably argue that insurance cover is an annually renewable contract which places no obligation on them to continue cover subject to the same terms and conditions, and because of this may impose new terms e.g., the level of policy excess. Specialist insurers can help, although they will be keen to understand causation and remedy before offering to provide cover.

If a Homeowner changes Insurer there is no obligation on the new Insurer to give subsidence cover, even if it existed under a previous policy with another Insurer. The new Insurer may reasonably argue that there is no obligation to accept a 'high risk' property but if a property has been properly dealt with, the risk should be non-existent or minimal. Much effort in recent times has been put into the problem of the transfer of insurance cover. Owners might also want to be cautious about changing Insurers if the possibility of a continuing risk exists, and be sure of the clarity of continued insurance cover. The ABI's "change of insurer" claims agreement[24] to which most insurers subscribe, significantly improves this situation.

If Insurers are prepared to pay for any reasonable repairs following damage and offer continued cover, the property should, in the vast majority of cases, continue to offer adequate security for a mortgage loan. However, this might depend on how any risk is perceived by a particular Lender from a marketability standpoint, and this in turn relates to the quality of investigation, diagnosis and remedy. A property offered as security for a mortgage loan needs to be regarded by a Lender as being reasonably readily saleable if it has to be taken into possession. Distortions in the structure and other signs of previous problems can influence this decision and some Lenders may not be willing to offer a mortgage on affected properties (Section 2.2). That said, many so-called 'period properties' are both distorted and at the same time are mortgageable. The ultimate test appears to be that of continued stability.

Where there is only nominal damage to the building and it remains substantially sound without any hazardous external features, the property should remain mortgageable provided the Insurers are willing to continue insurance cover, and such cover should not be unreasonably withheld or removed.

Subject to policy terms, Insurers will pay for remedial works but do not compensate the Insured for any notional loss of property value (usually referred to as 'blight'), which in any event, is a matter subject to various market forces. It is usual that any effect on property value, if it even exists, decreases with time.

4.7 Carbon management

The Climate Change Act 2008[29] commits the UK to a 100% reduction in emissions relative to 1990 levels by 2050 (i.e., 'net zero'). Interim targets have been introduced, challenging all industrial sectors to act quickly on reducing the carbon footprint of their activities. Insurance companies' responsibilities can therefore extend to ensuring that claims spend is managed in a sustainable way.

As far as subsidence claims are concerned there are areas where current practice has already developed to reduce carbon e.g., the reduction in the frequency of underpinning and greater focus on damage mitigation. However, the urgent need to act will require the development of sustainable subsidence strategies encompassing all aspects of risk and claims management, with possible areas of focus including:

- Risk management — claim prevention, including maintenance of drains and trees
- Investigation — operational approach
- More effective desk-based triaging
- Virtual investigations
- Remote monitoring
- Electric vehicle fleets
- Resource management
- Repairs
- Mitigation strategies
- Repair strategies — minimising strip-out (repair vs. replace)
- Appropriate use of crack stitching vs cosmetic repair
- Use of sustainably sourced materials
- Waste management approach
- Resource management (e.g., local trades)

5 Causes of damage

5.1 Introduction

This chapter focuses on the main causes of foundation and subsoil movement (and the associated technical aspects) that can result in structural damage:

- Settlement (Section 5.2.1)
- Consolidation (Section 5.2.2)
- Volumetric changes (clay soils) (Section 5.2.3)
- Volumetric changes (soils other than clay) (Section 5.2.4)
- Instability (Section 5.2.5)
- Erosion (Section 5.2.6)
- Mining (Section 5.2.7)
- Frost heave (Section 5.2.8)
- Vibration (Section 5.2.9)
- Chemical attack (Section 5.2.10)

Note however, that structural damage cannot always be attributed to foundation and subsoil movement. It can also result from a combination of other factors, including:

- Thermal movement (Section 5.3.1)
- Creep (Section 5.3.2)
- Moisture movements (Section 5.3.3)
- Roof spread (Section 5.3.4)
- Timber rot (Section 5.3.5)
- Masonry and brickwork damage (Section 5.3.6)
- Poor design, detailing and workmanship (Section 5.3.7)
- Fire (Section 5.3.8)
- Lack of maintenance (Section 5.3.9)
- Movement due to changes in load patterns (Section 5.3.10)
- Accident (Section 5.3.11)
- Calcium silicate bricks (Section 5.3.12)
- Bookend effect/thermal expansion (Section 5.3.13)
- Direct root action (Section 5.3.14)
- Differential settlement (not subsidence) (Section 5.3.15)
- Mundic concrete (Section 5.3.16)
- Lateral movement of foundations (Section 5.3.17)
- Subsidence of an adjoining building (Section 5.3.18)

This can make diagnosis a complex process.

5.2 Main causes of foundation and subsoil movement resulting in structural damage

5.2.1 Settlement

Insurance policies tend to define settlement as 'downward movement under applied load'. Provided the acceptable bearing pressure of the soil is not exceeded, the settlement should be small, often no more than a few millimetres, but bearing capacity should not be the only criterion. Determination of the likely settlement under the predicted load is a more important factor.

The effects of this initial settlement can be minimised by the inclusion of appropriate details to maintain the flexibility of services and, for example, later extensions to the structure. Any associated damage should manifest itself during the first few years after construction and should be easily repairable by redecoration.

For granular soils the settlement will normally occur almost at once and can be predicted from the knowledge of the deformation modulus of the soil. For clay soils however, there are additional long-term components to settlement which involve the gradual squeezing out of pore water as the soil comes to equilibrium. Hence, to obtain the total settlement of a structure founded on clay soil it is necessary to consider the effects of three separate components:

1. Elastic (undrained) settlement.
2. Consolidation (movements associated with the dissipation of pore water pressures).
3. Drained creep or secondary consolidation, which occurs at constant effective stress.

Elastic settlement occurs instantly as the load is applied and, using simple elastic theory, can be estimated from the undrained deformation modulus of the soil, the bearing pressure, the width of the foundation and empirical influence factors which account for the geometry of the foundation and the distribution of stress with depth.

Consolidation settlement is normally estimated from basic soil properties, e.g., the coefficient of consolidation, which can be measured in a standard oedometer test (BS 1377-2[30] and BS EN ISO 17892-5[31]) or can be estimated from the soil's liquid limit for normally consolidated soils.

The total consolidation settlement can then be found by adding the compression of each individual layer over the depth of soil affected. The change in effective stress for each layer can be estimated from tables which are based on elasticity theory.

The rate of secondary consolidation per log cycle of time is normally considered to be constant, so the total secondary consolidation of a stratum can be estimated from a creep coefficient based on long-term observations, depending on the design life of the building and the period over which the observations were made. These predicted secondary settlements are not as reliable as the primary settlement calculations and should be used with caution.

5.2.2 Consolidation

5.2.2.1 Fill

Made ground can be extremely variable unless there is careful selection, controlled placement and compaction of fill materials, and serious problems can arise if construction takes place on filled sites. Uncontrolled tipping can leave voids and organic matter in an unconsolidated condition, and further decay and deterioration leads to unequal and variable settlement characteristics, e.g., subsidence could occur on the first inundation. Buildings founded on unconsolidated and variable material are at obvious risk and this may be greatly increased by building across the borders between filled and virgin ground.

Decomposition of organic material within fill material can cause methane gas generation and combustion. These problems are most likely to occur in fills containing domestic refuse or industrial wastes (Figure 5.1). The presence of gas or other chemical contaminants may have health and safety implications.

Figure 5.1: Landfill site

The amount of self-weight consolidation of fill is dependent on the type of the original treatment and the material itself, as well as other external factors such as surcharge change in moisture content or vibration. A consolidation event can be triggered by one of these factors after a considerable period of time (100 years or more).

The change in moisture content of fill materials due to variations in groundwater tables or surface water collecting in the fill can bring about ground movement. This can be a particular problem with loosely backfilled ground. Another example is pulverised fuel ash (PFA) which has been known to lose its consistency and become unstable when saturated.

5.2.2.2 Vibration induced consolidation

Granular soils can be prone to consolidation by vibration and this may result in subsidence. It is generally the ground-borne vibrations caused by influences such as earthquakes and pile driving that can cause damage to buildings and their foundations.

When assessing vibration as a cause of structural damage, note that structures are far less sensitive than people. What may seem a serious vibration to a person may not cause damage to a building.

While vibration is rarely the sole cause of structural damage to buildings, it can cause cracking to plaster and (more rarely) cracking between structural loadbearing units, and may result in subsidence of granular soils. It is unusual for clay soils to settle under vibrating loads unless they have a high sand content.

Where vibro-compaction and vibro-consolidation techniques are used on new build projects to strengthen weak and compressible soils (particularly loose granular soils, silty or clay sands and silty soils) note that consolidation can also occur in soils beneath the foundations of nearby existing buildings and this may cause subsidence damage.

Where poor ground is consolidated, it is important to note that a building or structure can exert lateral forces beyond the limits of its footprint. It is essential that consolidation is undertaken to such limits that all lateral forces from the building have been sufficiently dissipated, so their residual components can be adequately resisted by the surrounding unconsolidated ground.

5.2.2.3 Secondary consolidation

After a building has experienced its initial settlement and the ground has undergone its primary consolidation, further consolidation may occur over a much longer period. Except in the case of peats, this secondary consolidation is rarely of any significant magnitude and can be ignored in most cases.

5.2.3 Volumetric changes (clay soils)

5.2.3.1 Shrinkage potential

Unlike granular soils, clays tend to shrink or swell as their moisture content changes. This is because the water is bonded to the surface of the clay particles, much as in a jelly, rather than simply filling the pores between the particles. The greater the proportion of clay minerals in the soil, particularly montmorillonite, the greater the potential for volume change. Areas of highly shrinkable clays in the South East of England are indicated in Figure 5.2.

The simplest way of obtaining an indication of the proportion of clay minerals and their ability to retain water within the clay structure of the soil is to measure its plasticity index, i.e., the range of moisture content over which the soil exhibits plastic properties. The upper plastic range is defined by the 'liquid limit' (i.e., the point at which the soil begins to flow) and the lower range by the 'plastic limit' (i.e., the point at which the soil stops being mouldable and begins to break up). As a very general guide, the higher the plasticity index the greater the shrinkage potential, but the correlation between these is poor over the range of changes in moisture content which occur within the field.

Under field conditions a soil is often not fully saturated, and moisture held within the soil structure can be released without volumetric change. Shrinkage will only occur if soil drying occurs within the clay matrix, and will diminish as the clay dries towards its plastic limit. Conversely, an increase in moisture content will cause the soil to swell, but in most over-consolidated clays there is a limit to the amount of water that will be absorbed under natural conditions. Three categories of volume change potential are given by the NHBC[32], but should be interpreted with caution for predicting actual shrinkage behaviour. BS 1377-2 provides further guidance. It should be noted that the NHBC Standards are intended to be used to ensure that new properties that come with a NHBC guarantee do not suffer from subsidence; it is not intended to be used to assess the potential subsidence damage caused by trees (Chapter 8 provides further guidance).

Figure 5.2: Areas of highly shrinkable clays in the South East of England

The moisture content of the clay can only change if the stresses acting on the soil change. There are effectively two ways in which this change of stress can be brought about:

1. External stress variation – a change in the imposed loading, e.g., the addition of foundation loads or the removal of the overburden.
2. Internal stress variation – the propensity for clay soils to absorb water, i.e., pore water suction is increased within the clay by evaporation (the atmosphere) and or extraction (vegetation), or alternatively pore water suction is decreased by the addition of water.

5.2.3.2 Desiccation

Desiccation of clay soils is often the cause of subsidence to buildings. The term 'desiccation' is often associated with the complete removal of moisture. When used in association with clay soils however, it means any reduction in the soil's moisture content associated with natural drying and/or extraction by vegetation.

In addition to the shrinkage potential of the soil, the volume change that occurs is dependent on the magnitude of the stress change. For most soils, there is a linear relationship between the volume change and the logarithm of the stress change. The increase in stress imposed on the soil by a low-rise building supported on strip footings or trench fill is modest, say 20–40kN/m^2 and the corresponding movement, i.e., settlement in firm clays, is in the order of a few millimetres. However the stress changes caused by desiccation are far greater, typically 300kN/m^2 and potentially as high as 1,400kN/m^2. These changes also tend to affect soil to a greater depth, particularly if there are trees or large shrubs present. It is quite conceivable therefore, that the movements associated with desiccation can be many times greater than those due to settlement under foundation loads, i.e., 100–200mm.

Research indicates that the effects of evaporation to the atmosphere in firm clay soils are largely restricted to the uppermost 0.5m of soil. Grass and other short-rooted vegetation can extend drying from 1.0–1.5m depth, and some large trees can extract moisture from depths of 5m or more. Additionally, the low permeability of high plasticity clays may prevent the moisture extracted during summer months from being replenished by rainfall and upward migration of the water table during the winter months. As a result, persistently desiccated soil develops under the tree which may increase in depth and lateral extent as the tree grows.

When a soil is described as desiccated it may in fact have a moisture content that is only 1–2% less than in its natural state, and even severe desiccation (Figure 5.3) is unlikely to reduce the moisture content of most UK clay soils by more than 10%.

Figure 5.3: Cracks in desiccated clay soil showing shrinkage effects

A reduction of the moisture content below the plastic limit allows air to enter the soil making it dry and friable, resulting in a pronounced effect on the soil's appearance. Near the surface, desiccation often causes the ground to crack and in very shrinkable soils this can be dramatic, with cracks 20–25mm wide and 0.5–0.75m deep in dry summer months.

It has been noted that London Clay with a normal moisture content just above the plastic limit of 30% near the soil surface, may fluctuate from 15% for dry summers to 40% for wet winters.

It has been known for a ground or basement level boiler room to dry out the clay stratum within its own vicinity and so cause its own subsidence. The addition of central heating to late 19th and early 20th century properties, some of which have shallow foundations, also causes drying out and shrinkage of the clay, in some cases leading to subsidence.

5.2.3.3 Effect of vegetation

The roots of all vegetation, whether grass, herbaceous plants, shrubs or trees, take moisture from the soil to replenish the water lost from their leaves (Figure 5.4).

During the summer, water uptake by vegetation is typically far greater than rainfall so that the soil is progressively drying (Figure 5.5).

Figure 5.4: Proximity of mature trees and bushes (e.g., Llelandii on left-hand side) to residential property may be a threat

Figure 5.5: Ground desiccation by tree roots

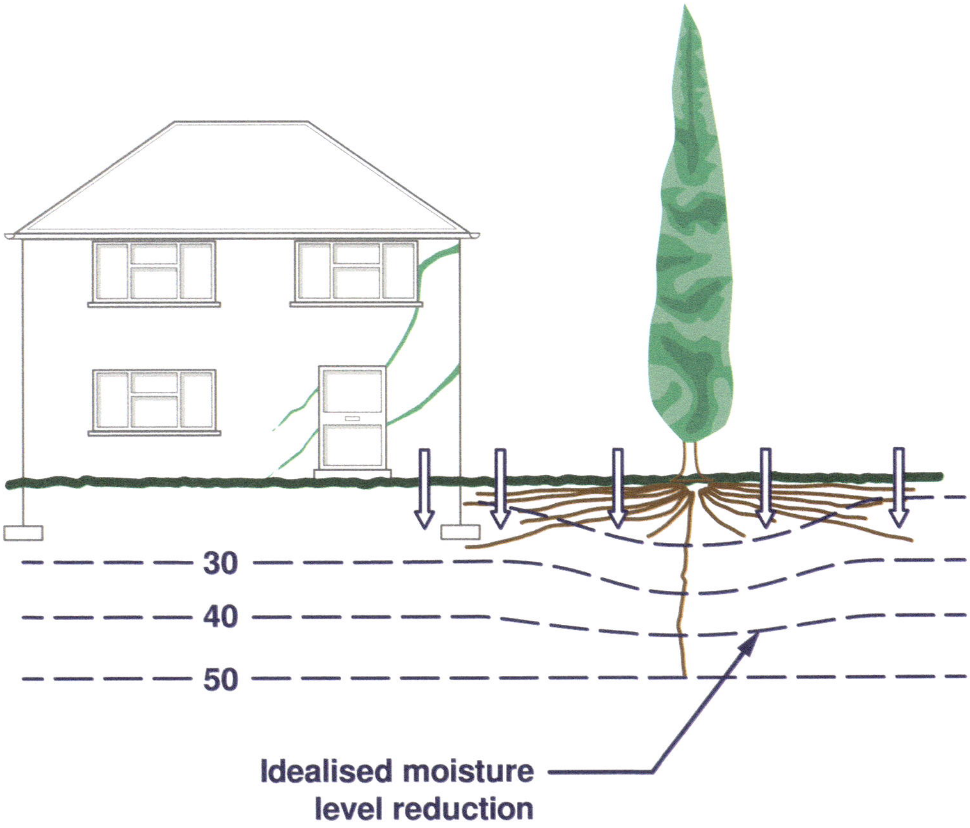

During the winter, deciduous plants lose their leaves and so have no water uptake; evergreen plants can continue some uptake, but the amounts are negligible compared to uptake in summer. As a result, during the winter rainfall is far greater than water uptake, so the soil rehydrates.

In clay soils, this produces a cycle of the soil drying and shrinking during the summer, and rehydrating and swelling during the winter. This can result in two types of soil movement:

1. Seasonal movements, in which the rehydration and swelling in the winter is the same as the drying and shrinkage in the summer, with the soil returning to its natural moisture content ('field capacity') by the spring. The amplitude of this movement will be increased in a dry summer.
2. In some situations, where the clay is of low permeability and soil-drying is occurring to sufficient depth, the moisture is unable to penetrate so that by spring there is still a deep-seated zone of desiccated soil. Over successive years, the depth and lateral extent of this zone can gradually increase. A persistent deficit of this sort typically develops during the period of rapid growth prior to maturity of a tree. Thereafter it can be maintained for many years through maturity. There may be a further increase in the persistent deficit in a period of prolonged drought. A single dry winter may create a semi-persistent deficit, but additional drying of this sort is rarely maintained, with the soil reverting to its previous persistent deficit in subsequent years.

If the tree which has created a persistent deficit is felled, instead of the soil being dried during the next summer, the process of rehydration will continue, with the soil progressively rehydrating and swelling, producing upward movement of the foundations ('heave'). The rate and duration of this process will depend on the permeability of the clay and the depth of desiccation. In extreme cases it can take at least 25 years, but more typically it is no more than a minor extension of seasonal drying with the process of rehydration complete in one or two years. Throughout this process, the rate of movement is gradually decreasing. Heave is no more than a long-term extension of the seasonal recovery. It follows that, whenever a tree has created a persistent deficit and a risk of heave, there will also be seasonal movement.

If a property has been built prior to the development of the persistent deficit, there will be pronounced distortion of the building caused by the soil shrinkage, and subsidence associated with the persistent deficit. This distortion will be corrected, at least partially, if the tree is felled and the persistent deficit dissipates. If a property is built on soil which is already desiccated, removal of the tree and resulting heave will lift the building to above its level at the time of construction.

It does not necessarily need a tree to be felled for a persistent deficit to diminish and heave to occur. Anything which causes a reduction in soil drying will mean that the recovery in winter is greater than the drying and subsidence in summer. For instance, pruning a tree which has created a persistent deficit will allow the soil to rehydrate and possibly heave, with a superimposed cycle of seasonal movement of reduced amplitude. Likewise, as a tree becomes overmature, its water uptake will decline.

The development of a persistent deficit requires a species of tree which is capable of causing drying of a clay soil to sufficient depth, e.g., poplar, oak, willow and cypress.

Development of a persistent deficit also requires a soil of low permeability. This is commonly associated with a soil of high plasticity. However, well-developed structural cracks in the soil, particularly if filled with sand or silt partings, will aid permeability. Larger features, whether natural such as bands of sand or gravel, or manmade such as backfill of drains, can aid rapid water movement and have a major influence in minimising soil drying — as can leaking drains or water mains.

5.2.4 Volumetric changes (soils other than clay)

Peat or soft silt may occur almost anywhere in the UK but is most likely in low lying fen, coastal, river estuary or valley and moorland areas. Perhaps the best way to regard peat is to think of it as reinforced water. It has high compressibility and may be displaced under load. Differential settlement can occur where peat layers are of a non-uniform nature and thickness and significant variations in depth may occur over very short distances.

Lowering of groundwater levels in peats, silts and loose granular soils whether permanent or temporary can bring about a significant reduction in the soil volume, often leading to subsidence. In fen areas where buildings are often constructed on natural surface crusts, drainage schemes can affect the stability of structures by removing moisture from underlying peat layers. Excavations for services can also have the same effect.

Due to its high compressibility, the secondary consolidation of peat can continue almost indefinitely[33], so its use for direct bearing of foundations to structures is not advised.

5.2.5 Instability

5.2.5.1 Slopes

The instability of sloping ground can result in sudden or gradual movement. The building design must take account of these movements to maintain structural stability. The assessment of slope stability requires specialised geotechnical expertise and this may be outside the experience of the general building professional. Natural clay slopes steeper than approx. 8° can suffer from relatively rapid instability. Long-term creep may occur on more shallow slopes. This is a complicated subject on which specialist advice is likely to be required.

5.2.5.2 Landslip

Landslip is a simple process whereby ground slides down to a lower level (Figure 5.6). An avalanche is a dramatic, rapid and extreme example of a landslip. Some landslips occur suddenly, and this most frequently happens after heavy rain or when frozen ground is thawing. Coastal erosion is another natural mechanism which can lead to landslip, also of a dramatic nature.

Figure 5.6: Landslip

When a landslip occurs in the vicinity of a building located near the top of a slope, the effect on the property is virtually identical to subsidence, namely the ground moves away from beneath the foundations of the building and the building effectively subsides.

Relatively shallow slopes can move over a period of time if the ground material is particularly susceptible to slipping.

5.2.5.3 Slope creep

Of greater relevance to most Homeowners is the potential for relatively shallow clay slopes (e.g., inclined up to approx. 8°) to migrate slowly downhill.

Damage caused by slope creep is commonly (and incorrectly) diagnosed as clay shrinkage, but it is possible that the two phenomena are related, e.g., desiccation cracks that open up in periods of exceptionally dry weather may not close completely. This fissuring (Figure 5.7) will allow water to enter the ground during wet weather thus encouraging rehydration and therefore heave. The water will lubricate the clay along the fissures thus encouraging movement downhill. Dehydrated ground will normally rehydrate, but ground that has moved downhill will not move back up.

Figure 5.7: Fissures can be an indication of slope instability

5.2.5.4 Excavations

Cutting away of sloping ground can likewise remove the support at the top or bottom of the slope and cause the ground to become unstable. Even temporary excavations for drains at the top or bottom of a slope can upset the balance.

Where the lateral stresses acting on clay soils are reduced, e.g., adjacent to a cutting, an excavation or a steep slope, the clay will tend to absorb water and soften. This process can reduce the bearing capacity of the soil and may in extreme cases lead to an increased possibility of foundation loads (or even the weight of the soil itself) inducing a shear failure in the soil. More commonly this process may produce some small lateral movement as it allows soil to swell and move towards a loosely back-filled trench, such as a deep drain trench. This softening will occur wherever the soil is below the water table for at least part of the year. In the UK this means that most excavations close to and below adjacent foundations are likely to induce some limited ground movements.

Having softened and thereby expanded to fill any available voids, the soil would again be at equilibrium and unlikely to be affected by a sudden influx of water, e.g., from a leaking drain. However, it is common practice to lay drains on a bed of granular material and in such circumstances, leaking water may erode the already softened clay.

5.2.5.5 Escape of water from drains

While an excess of water has little influence on clay soils below the water table, it may have a significant effect on desiccated soils. Hence, a long-term leak from a drain or water supply (Figure 5.8) can reduce the state of desiccation in the soil locally, producing recovery and heave.

Figure 5.8: Effect of broken or leaking drains

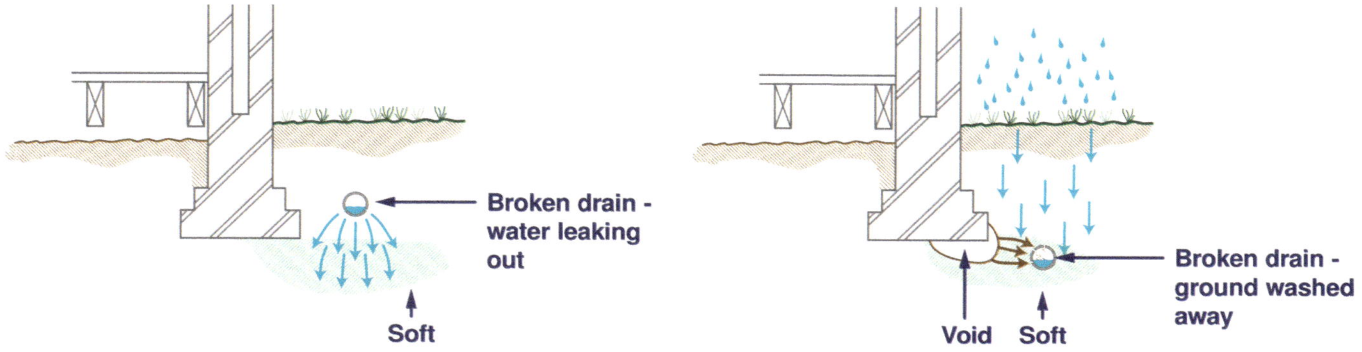

5.2.6 Erosion

5.2.6.1 Movement of fine soil particles

Internal erosion due to water seepage is a below ground level activity caused by groundwater transporting fine soil particles to another location e.g., where back-filled excavations exist or at the interface between different strata.

Ground movements due to water flow occur mainly in sandy or silty soils; the movement being caused by the removal of small particles which reduces the density of the remaining soil. The consequent loss of ground from beneath foundations may lead to subsidence or in extreme cases collapse of part of the structure. Ground movement by leaching or erosion in made ground or natural granular soil is characterised by a zone of loosened material.

Common causes of erosion due to water seepage are:

- Groundwater seeping into fractured sewers or culverts
- Carrying out excavations or groundwater pumping below the water table, without taking adequate precautions to prevent the loss of fines
- Seepage of water from leaking mains or drains which transport fine particles to a different location
- Positioning soakaways too close to the building
- Discharging of rainwater pipes into or onto the ground
- Water seepage causing solution of minerals from the ground, commonly known as 'leaching'
- Leaching of water and soils into backfill material placed around subsurface services, commonly known as the 'French drain effect'

Surface erosion may take place as a result of material loss in strong winds or by flowing water, e.g., floodwater. Wind is most likely to affect fine sands and silts on which external groundbearing elements are founded. Erosion can normally be prevented by encouraging the growth of vegetation, placing of paving or by other methods blanketing the erodable soil. Foundations at normal depths are not usually affected by surface erosion from wind.

Surface erosion by flowing water can be severe, particularly if structures are sited on slopes or at the bottom of valleys, where normal foundation depths of 900–1200mm can prove inadequate for cases of erosion by floodwater. Such erosion should be prevented by providing adequate drainage or other forms of surface protection, particularly in the paths taken by the periodical discharge of floodwater.

Unless they have a high sand/silt content, clay soils are unlikely to be affected by leaching or erosion since their low permeability prevents migration of the sand/silt content except by slow seepage. It is therefore unlikely that erosion would be the cause of subsidence in clay soils.

5.2.6.2 Natural cavities

The formation of natural cavities in the ground can result from a series of processes, but the majority are formed by dissolution, cambering and marine erosion. The action of both surface water and groundwater on soluble rock is the main cause of natural cavities in the UK. Natural cavities may be air-, water-, rock- or soil-infilled.

There is an inherent risk of subsidence when foundations are placed above these features. The most common cause of subsidence associated with natural cavities is the upward migration of the void, often caused by the destabilisation of the cavity infilling material. The commencement of subsidence often follows the introduction of surface water or changes in groundwater levels.

Natural cavities commonly occur in areas where there are soluble rocks such as chalk and limestone. These may be found at the surface or just beneath the surface strata. Mudstone rocks containing soluble salt or gypsum may be prone to the formation of natural cavities, as might areas where rocks have a soluble matrix or cement. They often occur where strong-jointed rocks such as sandstone and limestone overlie mudstone marl and clay. Fissures are formed in the rock by the cambering process, especially where there are steep slopes. The clay moves down the slope causing the overlying rock to fracture and form cavities.

In coastal areas natural cavities are formed where competent jointed rocks form a cliff line and become subject to marine erosion.

Stantec's Cavities Database[34] contains more than 32,000 UK records.

5.2.7 Mining

The effects of mining are discussed in detail in Appendix D. The physical removal of below-surface deposits can result in foundation subsidence. Mineral and rock deposits which have been mined in the past include coal, salt, limestone, tin and iron, etc. and historical records should be checked for evidence of these activities.

Both 'longwall' and 'pillar and stall' methods are described in Appendix D2. Longwall mining may distort and crack a property as the surface subsidence wave passes. The building may then be left in a reasonably stable condition. If the site is situated on the edge of the mining area, resulting differences in level can be unacceptable.

Shallow pillar and stall workings do not normally have an immediate effect on foundations. However, deterioration of the pillars, and outside influences such as groundwater can lead to the collapse of the working and serious surface damage many years after the causative mining (Figure 5.9).

Figure 5.9: Effects of shallow pillar and stall mining

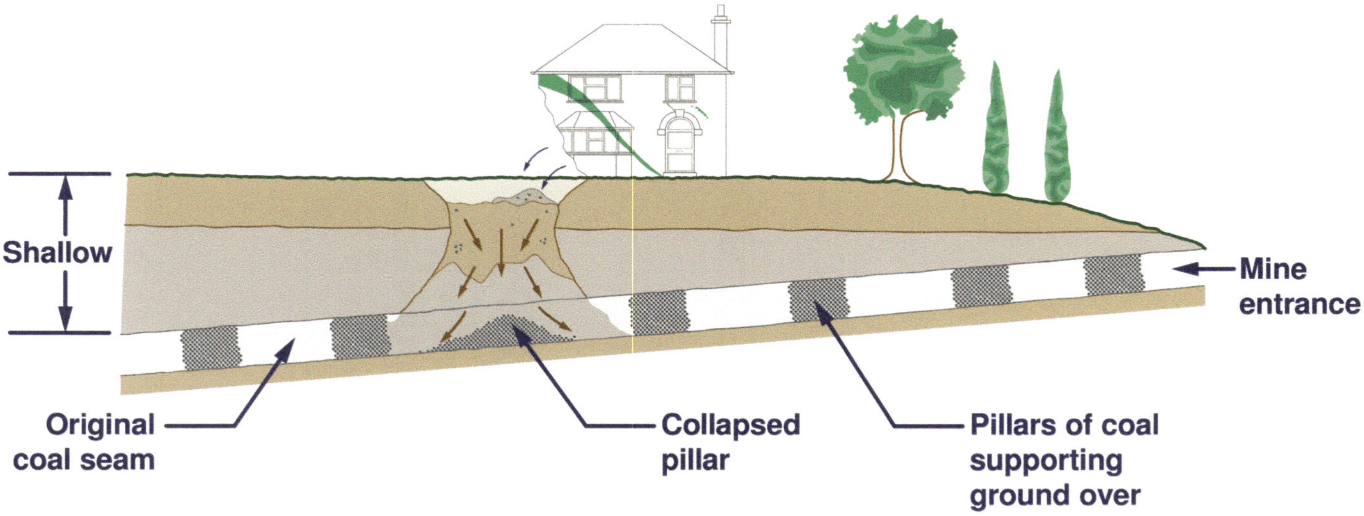

Reinstated opencast mining should be considered as a backfilled site, but may encounter other issues, e.g., deterioration of mudstone fill in water. Such sites can often have a considerable depth of fill.

Metalliferous mining is outlined in Appendix D3.

5.2.8 Frost heave

Soils that are subject to expansion when frozen can include silts, chalk, silty sands and some clays. Winter freezing of ground in the UK is normally confined to depths of less than 0.5m. During severe and prolonged periods of cold weather, freezing may affect the ground to depths approaching 1m. In these extreme conditions it was previously considered that the ground in the vicinity of a heated building is most unlikely to freeze because of the heat stored in the ground under and adjacent to the building. Modern houses are constructed with insulation under the ground floor and therefore the frozen ground may be closer to the building than previously considered.

Unheated buildings on frost-susceptible soils in areas where severe winter conditions are likely are of most concern, and precautions may be necessary.

Inadequate insulation where refrigeration or cryogenic plant is close to the soil can lead to localised frost heave.

5.2.9 Vibration

Vibration causes particles to accelerate from a static position. They reach a peak velocity and then displace before recovery. Further cycles of movement follow until the energy decays or is dampened. The maximum velocity (peak particle velocity) is expressed in mm/s and is often used as a criterion for damage to buildings. It can also be used to determine comfort levels for occupants of buildings.

The peak particle velocities are based on frequencies of 5–50Hz, which is the range most commonly encountered in buildings. The level of human perception of vibration occurs when the peak particle velocity is in the order of 0.3mm/s, the vibration becomes annoying at 2.5mm/s and unpleasant or even painful on reaching 5mm/s at the higher frequencies. Damage to decorations, plaster cracking, etc. might be expected to occur when the peak particle velocity is 2.5–7mm/s at the higher frequencies and 20mm/s at the lower frequencies, i.e., 5Hz.

Vibrations may be airborne, i.e., caused by wind, which except in the case of explosions will normally result in no more than door or window rattle, and even in extreme cases little or no structural damage to well-designed buildings in good condition.

There are several causes of groundborne vibration which can cause damage to buildings and their foundations:

- Road traffic-induced vibration is usually the result of a poor road surface, e.g., potholes
- Trains create larger groundborne vibrations in their immediate vicinity, which lead to complaints by Owners and occupiers of nearby buildings. Recorded velocities indicate that while levels of discomfort can be high, very few are likely to cause damage to buildings, other than minor cracking of plaster, etc.
- Pile driving causes shock waves that decay quickly in soft ground, but can be refracted or reflected in firm/hard ground. The structural condition of the surrounding properties may be affected, with the risk of damage diminishing with distance. While operators have been successfully sued, improvements in pile driving techniques have considerably reduced the risk of damage to adjacent structures

Nevertheless, an element of risk still exists which can be minimised by taking adequate precautions and assessing the risk of damage before work begins. CIRIA Technical Note 142[35] provides details of groundborne vibrations from piling. Further useful information is given in BS 5228-2[36] and BS 6472-1[37].

Less common causes of vibrations that may affect low-rise buildings include:

- Out-of-balance machinery
- Reciprocating engines
- Rock blasting
- Earthquakes
- Explosions

Vibrations caused by machinery are usually dampened by using special mountings to prevent damage and nuisance to occupants and buildings.

If structural damage is suspected, expert advice should be sought at an early stage. This will help limit or prevent further damage, particularly if the structure is of unsound condition.

A schedule of condition carried out on adjoining buildings before using vibration techniques can help avoid disputes, by identifying existing damage and the likelihood of a building being sensitive to vibration.

5.2.10 Chemical attack

The expansive reaction between chemicals in certain soils in contact with concrete can cause foundation movement. Some fill materials historically used underneath ground floor slabs may also be subject to various expansive chemical reactions which can result in the lifting of slabs and the walls they support. Heave of floor slabs supported on sulphate-rich fill material is particularly common. On rare occasions sulphate attack can weaken foundation elements such as piles and result in failure.

Sulphate attack on concrete is one of the most common forms of chemical attack and is addressed in the BRE's *Concrete in aggressive ground*[38]. Sulphate attack can only occur when water is present. Aqueous solutions of sulphate attack the hardened concrete causing it to expand. Sulphates occur naturally in some soils, e.g., London Clay, Lower Lias, Oxford Clay, Kimmeridge Clay and Keuper Marl. Additionally, some backfill materials such as colliery shale, brick rubble with adhering plaster and some industrial and mine wastes are rich in sulphates.

A high or fluctuating water table will normally saturate the whole of the foundations and could cause general uplift from the chemical attack, but this is known to be very rare. However, where the source of water is internal leakage or external run-off, the damage may be concentrated in the wettest areas.

Sulphate attack penetrates concrete slowly, and massive concrete is therefore less susceptible to damage than thin sections. The rate of attack is increased if moisture is lost via evaporation or leakage from one or more of the concrete surfaces and continuously replenished by another. Basement culverts, retaining walls and ground floor slabs without damp-proof membranes are more vulnerable than piers, trench foundations or piles. The risk of attack can be reduced by ensuring that the concrete has low permeability, i.e., using a low free water:cement ratio in the mix and/or by using a sulphate-resisting cement. Methods of testing soils for sulphate content can be found in BS 1377-3[39].

Chemical reactions within fill beneath a building can cause expansion of the fill material. Unhydrated lime or magnesia will produce volume expansion in the presence of moisture. Fill materials that contain a significant amount of clay, e.g., colliery spoil, can swell if they are placed in a dry condition and then subsequently become wet.

Materials that contain old blast furnace slag or broken concrete may be affected by sulphate expansion when in contact with groundwater. Recently-produced slag is not generally susceptible to sulphate attack, but all such materials should be checked before use.

In the presence of air and moisture, sulphides in the form of pyrites can oxidise to form soluble sulphate. With pyrites in the presence of calcite there is a danger of expansion due to the growth of gypsum crystals.

A relatively small number of cases of thaumasite sulphate attack (TSA) were first discovered in buried concrete in the 1990s. TSA will only occur in buried concrete when all the primary risk factors are present simultaneously and developed to a significant degree. These primary risk factors are:

- Presence of sulphates and/or sulphides in the ground
- Presence of mobile groundwater
- Presence of carbonate, generally in the concrete aggregates
- Low temperatures (generally below 15°C)

Additionally, there are a number of secondary factors which influence the occurrence and severity of TSA and its effects:

- Type and quality of cement used in concrete
- Quality of concrete
- Changes to ground chemistry and water regime resulting from construction
- Type, depth and geometry of buried concrete

Advanced TSA is easy to identify visually as it typically exhibits a white pulpy mass at the surface of the affected buried concrete. TSA in its earlier stages through to more advanced stages may be identified in a laboratory using various analytical techniques.

The structural consequences of TSA are rarely of serious concern in respect of public safety, as most buildings and structures would exhibit warning signs of distress.

5.3 Other factors causing structural damage

5.3.1 Thermal movement

The expansion and contraction of differing building materials without the provision for such movement may cause cracking or distortion (Figure 5.10). A further example includes brick facing panels to concrete frames that have no allowance for movement.

Figure 5.10: Cracking due to thermal movement

5.3.2 Creep

Building materials can deform slowly under load. This can result in fracture and distortion. Timber components such as floor joists and beams are particularly susceptible to this effect. A further example is where reinforced concrete frames compress under load over a period of time, but the enclosed brickwork facing panels, where constrained between the floor slabs, bow outwards due to the reduction of the storey height.

5.3.3 Moisture movements

The drying out of materials, especially in new houses, can cause noticeable cracks. Cracks can also occur in older properties following the introduction of central heating, which can cause drying shrinkage. Calcium silicate bricks (Section 5.3.12) and some other concrete masonry products are particularly prone to initial shrinkage cracking.

Movement joints in walls formed from susceptible materials, such as concrete blocks, can limit shrinkage cracking. The initial moisture expansion of new clay bricks can be substantial and is an irreversible chemical process unlike normal reversible moisture movement. This is usually only an aesthetic problem but in some instances can be sufficiently serious to warrant its consideration at the design stage, e.g., in long panels of brickwork.

5.3.4 Roof spread and movement

Inadequately tied pitch roofs can spread quite appreciably towards the eaves under permanent, imposed and wind actions, causing external walls to distort and sometimes crack (Figure 5.11a-b).

Figure 5.11a Roof spread (non-subsidence) causing distortion and cracking

Figure 5.11b Lack of roof bracing causing lateral movement and cracking to brickwork below

5.3.5 Timber rot

The deterioration of heavily loaded timber members supporting part of the building may result in instability or damage e.g., rot (Figure 5.12).

Figure 5.12: Timber beam severely rotted (non-subsidence)

5.3.6 Masonry and brickwork damage

The formation of sulphates in mortar can occasionally cause walls to crack. Old coal-fired chimneys are susceptible to this effect. Severe frosts have occasionally caused the expansion of moisture-filled external cracks (Figure 5.13). The deterioration and rusting of embedded metal, such as old wall ties, is more common and can also cause masonry damage (Figure 5.14).

Figure 5.13: Frost damage to brickwork (non-subsidence)

Figure 5.14: Crack along lintel due to rusting of embedded steelwork (non-subsidence)

5.3.7 Poor design, detailing and workmanship

Inferior design and detailing and the use of poor quality materials and workmanship may eventually bring about cracking (Figure 5.15).

Figure 5.15: Brickwork built on (subsequently rotted) window frame

5.3.8 Fire
Severe fires can cause expansion, cracking, bulging, distortion and spalling of structures. Fire-induced cracks (often mistaken for subsidence) can occur in the outside wall behind a chimney flue.

5.3.9 Lack of maintenance
The saturation of masonry due to leaking gutters or poor pointing/rendering can cause delamination and cracking especially when combined with frost action (Figure 5.16).

Figure 5.16: Cracking in render due to moisture penetration

5.3.10 Movement due to changes in load patterns
Cracking can occur from redistribution or increases in the loading of a structure, e.g., if structural alterations have been carried out without proper appraisal (Figure 5.17), or if a heavy material is stored in part of the building.

Figure 5.17: Pointing repair of brickwork cracks following sagging of lintel at ground floor ceiling level (non-subsidence)

5.3.11 Accident

Accidents through influences such as extreme weather, explosion or impact can be the cause of structural damage (Figure 5.18).

Figure 5.18: Cracking due to vehicle impact damage (non-subsidence)

5.3.12 Calcium silicate bricks

These bricks have a predominant characteristic of contraction, which will tend to occur if they are saturated before laying and subsequently dry out. This is an irreversible effect with the result that unrestrained brickwork will change size and crack if inadequate movement joints are used.

5.3.13 Bookend effect/thermal expansion

Some older terraced properties — where the internal party walls lean away from the centre of the terrace towards the ends — have been known to show signs of movement, with the consequence of the end of terrace gable walls being substantially deflected. Such distortion can be accompanied by lateral and diagonal cracking to the front and rear elevations.

5.3.14 Direct root action

The growth of trees and root systems in very close proximity to structures can result in distortion and damage. The forces exerted by growth of the base of the trunk, and roots near the surface, are comparatively small compared to the weight of a house and damage would normally be limited to lightweight structures such as low, freestanding walls and paths (Figures 5.19 and 5.20).

Figure 5.19: Cracking due to growth of tree roots

Figure 5.20: Direct root action damage to wall

The greatest risk of direct damage occurs close to the tree from the growth of the main trunk and the roots. This damage will diminish with distance and the risk of direct damage can be minimised by taking precautions during construction. BS 5837[41] gives recommendations for the minimum distance between the structure and the trunk. Maintaining these distances should avoid direct damage to a structure (including foundations) from the future growth of the trunk and roots.

When buildings are within the distances recommended in BS 5837, precautions should be taken to allow for future growth. These precautions may involve reinforcing foundations to resist lateral thrust, and bridging systems to allow sufficient clearance.

Localised root growth can be encouraged by water leaking from drains. If there is a pre-existing defect in a drain, a root is liable to enter and then proliferate within the drain, often leading to blockage. Further expansive growth of a root may be able to enlarge the defect, but will not initiate damage to an intact drain.

5.3.15 Differential settlement (not subsidence)

Foundations constructed at different depths may settle at different rates. Common situations that can induce differential settlement include basements under only part of the building footprint (Figure 5.21), building extensions, and buildings that have only been partially underpinned.

Figure 5.21: Severe cracking due to wall on shallow foundations adjacent to wall over basement (not subsidence)

5.3.16 Mundic concrete

Concrete that uses mining waste as aggregate can suffer serious deterioration, particularly where moisture and oxygen from the atmosphere are also available. The generic name given to this class of aggregate is 'mundic' derived from the Cornish word for iron pyrites (FeS_2), although the term is now used to cover a group of waste product aggregates of this type.

The problem was first identified in a significant proportion of buildings in Devon and Cornwall, although it also exists to a lesser extent in other parts of the UK. The Cornish experience suggests that the problem, where present, is most likely to affect domestic and similar low-rise buildings built pre-1950 using concrete or concrete products incorporating locally available mundic aggregates.

The problem has affected some types of manufactured concrete blocks in particular, but also some cast *in situ* concrete where mundic has been used as aggregate. It mainly affects the superstructure of buildings (particularly external walls built in concrete blocks) rather than foundations. The evidence of deterioration frequently displays a chequerboard crazing on surface finishes such as rendering and plaster, or on the exposed faces of *in situ* concrete. The RICS has published a useful guide[42].

5.3.17 Lateral movement of foundations

There is an assumption that foundations will only move in the vertical plane, but they can also move laterally e.g., at a change of level. However, there may be hidden features, such as a below ground storage tank, which can compromise the lateral soil strength, so that the bulb of soil pressure under the foundation pushes the soil laterally. Figure 5.22 show a situation where the base of the wall has moved laterally as well as downwards. The only indication of a 2m deep water run-off storage tank was an inspection cover.

Figure 5.22: Lateral movement at base of wall due to lateral movement of foundations

5.3.18 Subsidence of an adjoining building

Where an otherwise stable property is damaged as a result of activity on an adjoining site, e.g., subsidence or trees, the Owner of the stable property must reasonably have a claim against their neighbour (Appendix C). This may be covered by the neighbour's third-party insurance.

6 Initial appraisal

6.1 Introduction

The primary objective of the initial appraisal is to determine whether the reported defects to a property meet the damage threshold and qualify for further investigation. Experience of subsidence investigation and repair works provides excellent background to carrying out an initial appraisal which will inform further investigations and minimise the risk of unnecessary investigative works.

In this chapter, the use of rule of thumb and other preliminary investigative techniques (including the desktop study) are discussed, as are the characteristics of potential causes of subsidence, including: the property's age and condition; the proximity of trees and other vegetation; geotechnical features and drainage. Consideration of these factors can save both time and resources when undertaking a subsidence-related investigation. The chapter concludes by giving an indication of what a typical initial appraisal report should contain.

Concerns around possible subsidence are usually raised when:

- The Owner considers that the damage may relate to subsidence and contacts their insurance company in line with the policy conditions. The Insurer will then usually appoint their own Expert to inspect the damage
- The Owner, having identified cracks, personally consults an Expert to inspect and advise on the damage observed. At this stage the Owner may or may not suspect that the damage is associated with subsidence, and indeed it may not be
- The Owner may suspect that damage has occurred because of subsidence and therefore notify their Warranty Provider (if the building is typically less than 10 years old), Insurer, or Lender if there is still an outstanding mortgage. At this stage, the Warranty Provider or Insurer will either directly appoint a suitable Expert to carry out an 'initial appraisal' or may advise the Owner to do so. It is usual for the Insurer to make the appointment, or at least approve the Owner's choice of Expert
- Immediately prior to advertising a property for sale, damage to the building fabric may be identified by the Estate Agent when advising the vendor on the marketability of the property. The damage may be identified by a Valuation Surveyor during a sale when acting for a prospective purchaser or Lender. The Lender will then decide what further action may be required

Once the possibility of subsidence has been raised, the damage may require further investigation (Figure 1.3).

6.2 Objectives

Irrespective of the initial concern, the principal objectives of the initial appraisal are:

- To make a preliminary assessment of the significance of the damage
- To consider if the damage may be due to foundation movement (i.e., heave or subsidence) or some other cause which may or may not be an insured peril covered under the policy
- (if the damage is thought to be due to foundation movement) To consider the possible mechanism of that movement
- To consider whether the cause of the movement can be resolved at this stage to achieve stability, although this may be deferred until further investigation
- To consider whether repairing the superstructure is sufficient, or whether a more detailed investigation is required
- To identify the requirement for further investigations. Careful appraisal at this stage could significantly improve the quality of any subsequent investigation work
- To consider the scope of possible repairs
- To allow Insurers or Warranty Providers to give an indication of their liability, and in doing so, relieve the burden/worry of the Owner

Where the Expert is appropriately experienced or has been specifically briefed by the Warranty Provider or Insurer, the Expert may be required to comment on whether there is a possible breach of any of the policy conditions which might require further investigation, or alternatively the operation of one of the policy exclusions.

6.3 Suitable professional Experts

An Expert must be adequately experienced and appropriately qualified. Those likely to be competent to undertake an initial appraisal are:

- Chartered or Incorporated Engineers
- Chartered or Incorporated Valuers & Surveyors
- Chartered Builders
- Chartered Loss Adjusters
- Claims Investigators or Specialists
- Arboricultural Consultants

More detail is provided in Chapter 2.

Not all those who undertake an initial appraisal will be eligible to proceed with the more detailed investigation into the cause of the problem and be able to consider alternative solutions as outlined in Chapters 7–9. If, after the initial appraisal, it is decided that a more detailed investigation is required, this is best undertaken by a Chartered Structural or Civil Engineer with appropriate experience.

The use of video technology is increasingly becoming used for the initial visit. In some instances, and particularly on less complex claims, this involves a non-expert attending the property to video and photograph the damage, capture all pertinent information and upload this information online. It is then assessed by the Expert remotely and any further queries raised with the Owner. Another option is for the Expert to arrange a virtual meeting with the Homeowner. Here, the Expert can direct the Owner to live feed the areas of damage and concern. This technique is useful for Loss Adjusters and Insurers in times of significant upturn in claim numbers; enabling the Expert to have increased capacity to assess damage on a larger number of properties. It is important that this technique is used in conjunction with suitable triage to ensure its use on the right profile of property/damage/claim.

6.4 Procedure

6.4.1 Desktop study

Prior to attending the property/site a desktop study should be undertaken. There are many sources of information available to help compile the study:

Online maps provide satellite images which can reveal locations of trees, water courses, ground slopes and drainage manhole locations.

Images of the property at street level show the front of the property but may also show how the property looked in previous years. This can be useful for understanding whether any vegetation has grown/been removed in recent years.

The British Geological Survey's Geology Viewer[43] contains digital surface geological maps of Great Britain (Figure 6.1). This provides information on the likely subsoil conditions, although ground investigations should always be undertaken to confirm the actual conditions.

The Planning Portal[44] **and some Local Authority websites** may provide both historic and recent planning and building regulation applications for the property. This has the benefit of possibly providing floor plans which will assist in marking up the damage. This could also show any structural alterations to the property.

Estate agencies often retain information such as photographs and floor plans of recently sold properties. This can provide useful information regarding the condition of the property prior to a previous sale.

A request for **survey reports** should be made to the Owner to establish the condition of the property and any structural concerns raised by the previous survey/s.

Geo-mapping data providing geological information on the site itself and/or the immediate vicinity may be available. This can help with background information relating to previous claims, investigations or soil data.

Figure 6.1: British Geological Survey's Geology Viewer

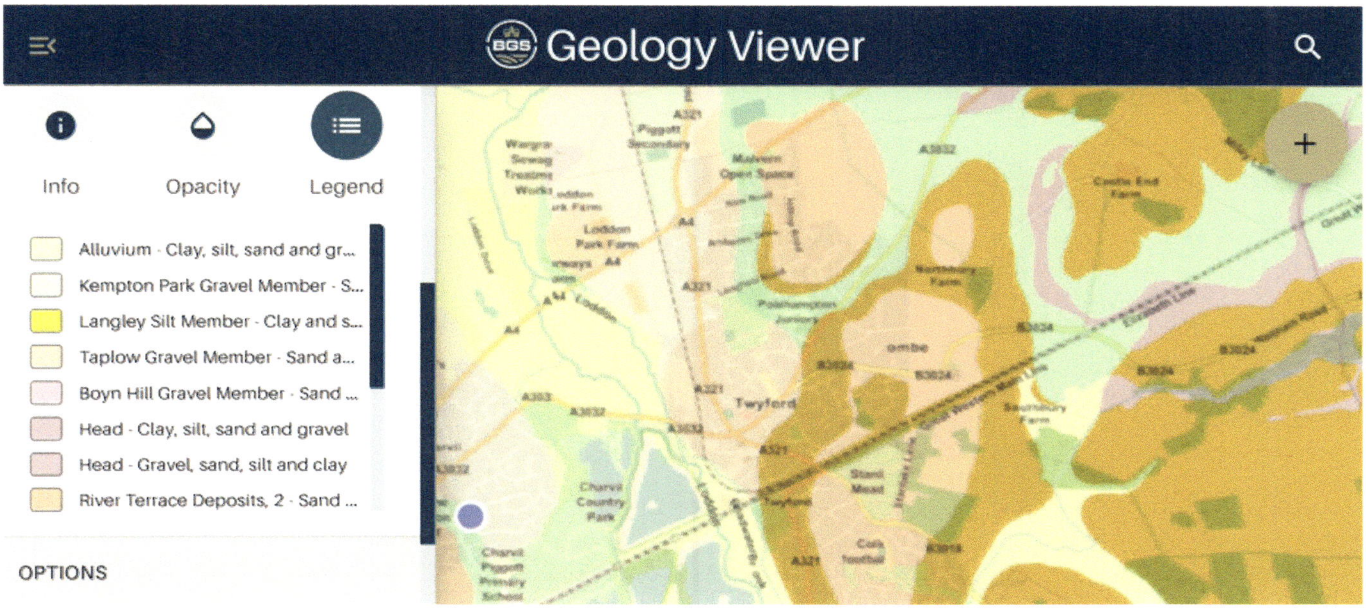

6.4.2 Fundamental considerations

It is incumbent on the Expert when carrying out and reporting on the initial appraisal that the views given are balanced and a fair presentation of the facts at hand.

The Expert should appraise the surrounding area to identify any geological features (water courses, sloping ground, old mine workings), building work being carried out on adjacent land and the presence of any large trees or other vegetation.

The Expert should discuss the history of the property with the Owner, including any structural modifications and/or extensions and when the damage first occurred.

It may not be possible to diagnose the precise cause of subsidence at this stage, but in the vast majority of cases, principal subsidence criteria can be identified. The more important criteria to consider are:

- The age of the property and any alterations (Section 6.4.2.1)
- Damage threshold and the significance of cracks (Section 6.4.2.2)
- The timing of the damage (Section 6.4.2.3)
- The proximity of the property to trees or large woody shrubs (Section 6.4.2.4)
- The proximity of the property to sloping ground or geo-features (Section 6.4.2.5)
- The likelihood of leaking drains (Section 6.4.2.6)
- The likelihood of the property being on subsidence-sensitive soils (Section 6.4.2.7)

6.4.2.1 The age of the property and any alterations

New builds are 'untested' and it is therefore necessary to adopt a particularly cautious attitude if cracks or other evidence of structural movement are found to exist within the fabric. Initial settlement affects almost all properties after construction, but damage resulting from initial settlement should never give rise to anything more than cracks of hairline or slightly greater size. Cracks of more than 1mm in width, especially if there are a number of cracks in any one wall should be scrutinised. Cracks of more than 5mm are potentially serious.

For the purpose of the initial appraisal, a 'new build' should be defined as 'built within the last five years'. Younger properties which display evidence of cracks of 1mm or more in width or other structural movement should be referred to an Expert for further investigation. This is particularly necessary if the property was redecorated a year or so after construction and cracks have developed afterwards.

It is highly likely that a privately-owned domestic property less than five years old will have the benefit of an insurance-backed warranty scheme and, irrespective of who identifies the damage, the Owner is advised to report the existence of the damage to the Warranty Provider.

Failure to report evidence of major structural movement once it is observed, and certainly within the first 10 years after construction, can mean that the Owner is likely to have no recourse through the warranty.

Although the building's Insurer (and Lender if involved) will need to know promptly of the existence of structural damage, they will primarily wish to know whether the damage has significantly affected the value of the property (in the case of the Lender) or whether it results from an insured risk (in the case of the Insurer).

It may be that the problem has arisen from an uninsured risk and so responsibility for rectifying the damage will fall to the Owner without assistance from the Insurer.

In most cases it is reasonable to assume that movements due to initial settlement will have ceased in properties that are more than five years old, and that inherent defects likely to cause subsidence will have manifested themselves and been discovered, e.g., poor construction, leaking drains, soft spots in the ground.

However, for properties that are at least five years old, the Expert undertaking the initial appraisal should consider the following six factors (Sections 6.4.2.2–6.4.2.7) in the order in which they are presented.

6.4.2.2 Damage threshold and the significance of cracks

On occasions buildings may be subjected to differential ground movements. These movements can result in damage which is of concern and which is therefore considered to be above the 'damage threshold' or 'serviceability limit'. These are the cases of subsidence which definitely require more detailed consideration and need to be subjected to an initial appraisal.

In the early years of subsidence claims investigations, an acceptable level of cracking (damage threshold), was not defined. Minor cracks resulting from trivial movements, however caused, were attended to by Owners as part of on-going maintenance. More recently though, even minor cracking has raised Owner concerns, not necessarily for aesthetic reasons, but more on account of the anxiety generated because the true structural significance of damage was not appreciated by interested parties. This became most detrimental on property values due to the nervousness of Valuation Surveyors, Lenders or Insurers — because most property sales were invariably relying on one or more of these parties for sales to be achieved.

It is also fair to say that, because of the increase in personal expectations of what people perceive to be their entitlement from a property, the tolerance level of occupants with regard to minor cracks has decreased sharply. At the same time, awareness of subsidence increased — with the result that cracks which were previously ignored began to assume disproportionate significance.

However, Owners not contemplating selling their properties are often considerably more tolerant of minor cracking once it has been explained that the cracks are not defects likely to affect the serviceability of their property. Some cracks, if they are static, can be tolerated when they are out of sight e.g., cracking damage within the roof space caused by point loading from a roof support.

The important questions are:

- When should cracks and other evidence of structural movement be considered significant?
- Is the defect caused by non-structural effects like drying shrinkage of plasterwork or settlement of a timber framed structure, which should be anticipated and catered for?
- Does the damage result from subsidence?
- How do Owners assess the damage or seek advice?

Table 6.1 uses BRE classifications[6] to offer descriptions (and suggestions for repair) of wall cracks of varying severity. Table 6.2 uses BRE classifications[6] to describe cracking caused by ground floor slab settlement.

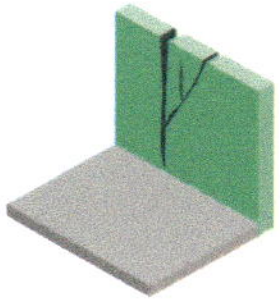

Table 6.1: Classification of visible damage to walls

Category	Approx. crack width	Description of damage	Repair types/considerations
		Cracks defined as **hairline**	
0	<0.1mm Up to 0.1mm	Generally considered to have negligible structural implications, and can be expected to occur in almost all buildings at any location. They are not generally related to subsidence/foundation movement[45].	Internally, cracks can be filled or covered by wall covering, and redecorated. Externally, cracks rarely visible and remedial works rarely justified.
		Cracks defined as **fine** (Figure 6.2)	
1	<1mm 0.1–1mm	These cracks may occasionally have some structural significance, but are not usually deemed serious. Often these cracks are more visible inside buildings than in external brickwork. Would generally be located at points of structural weakness, e.g., window/door openings. Indicates slight foundation movement, particularly if isolated. An array/series or large number of closely located fine cracks is unusual, but could signify more substantial foundation movement.	Internally, cracks can be filled or covered by wall covering, and redecorated. Externally, cracks may be visible, sometimes repairs required for weather tightness or aesthetics. Note: Plaster cracks may reappear if not covered by a wall covering.
		Cracks defined as **moderate** (Figure 6.3)	
2	<5mm 1–5mm	These cracks are likely to have some structural significance and will almost always occur at points of weakness or hinge points. Generally cracks will be visible internally and externally and will indicate foundation or other structural movement sufficient to distort door and window frames and make doors and windows stick. Weather tightness could be an issue that needs to be investigated as could the structural integrity of the building.	Internal cracks are likely to need raking out and repairing to a recognised specification. May need to be chopped back, and repaired with expanded metal/plaster, then redecorated. The crack will inevitably become visible again in time if these measures are not carried out. External cracks will require raking out and repointing. Cracked bricks may require replacement.
		Cracks defined as **serious** (Figure 6.4)	
3	>5mm <15mm 5–15mm	There will almost certainly be some compromise of the integrity of the structure and weather tightness may be impaired. Serious distortion to door and window frames may occur, and glass fracturing is possible, as are service fractures and strains.	Internal cracks should be repaired as per Category 2, plus perhaps reconstruction if seriously cracked. Rebonding will be required. External cracks may require reconstruction perhaps to panels of brickwork. Alternatively, specialist resin bonding techniques may need to be employed and/or joint reinforcement.

(continued)

Table 6.1: *continued*

Category	Approx. crack width	Description of damage	Repair types/considerations
		Cracks defined as **severe** (Figure 6.5)	
4	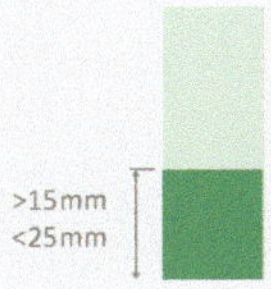 15–25mm	Structural integrity severely compromised — floors sloping; walls leaning or bulging; bearings of beams; lintels suspect. Pipe fractures and straining likely — windows broken.	Major reconstruction works to both internal and external wall skins are likely to be required. Realignment of windows and doors may be necessary.
		Cracks defined as **very severe** (Figure 6.6)	
5	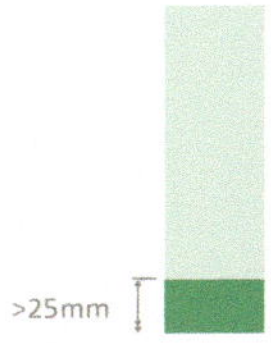 Greater than **25**mm	Potential danger from failed or fractured structural elements. Safety issues must be considered.	Major reconstruction works, plus possibly structural lifting or sectional demolition and rebuild may need to be considered. Replacement of windows and doors, plus other structural elements possibly necessary. Note: Building and CDM Regulations will probably apply (Sections 10.4, 10.6 and Appendix E).

Figure 6.2: Typical Category 1 crack

Figure 6.3: Typical Category 2 crack

Figure 6.4: Typical Category 3 crack

Figure 6.5: Typical Category 4 crack

Figure 6.6: Typical Category 5 crack

Table 6.2: Classification of visible damage caused by ground floor slab settlement

Category	Approx. crack width	Approx. gap*	Description of typical damage
0	N/A	Up to 1mm	Hairline cracks between floor and skirtings.
1	N/A	1–6mm	Settlement of the floor slab, either at a corner or along a short wall, or possibly uniformly, such that a gap opens up below skirting boards which can be masked by resetting skirting boards. No cracks in walls. No cracks in floor slab, although there may be negligible cracks in floor screed and finish. Slab reasonably level.
2	Up to 1mm	6–13mm	Larger gaps below skirting boards, some obvious but limited local settlement leading to a slight slope of floor slab; gaps can be masked by resetting skirting boards and some local rescreeding may be necessary. Fine cracks appear in internal partition walls which need some re-decoration; slight distortion in door frames so some 'jamming' may occur, necessitating adjustment of doors. No cracks in floor slab although there may be very slight cracks in floor screed and finish. Slab reasonably level.

(continued)

Table 6.2: *continued*

Category	Approx. crack width		Approx. gap*		Description of typical damage
3	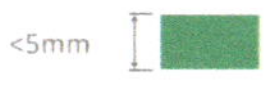	1–5mm	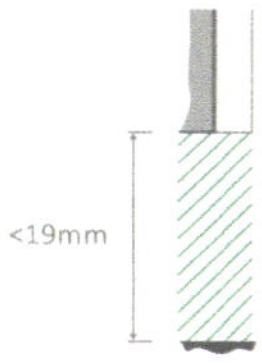	13–19mm	Significant gaps below skirting boards, with areas of floor (especially at corners or ends) where local settlements may have caused slight cracking of floor slab. Sloping of floor in these areas is clearly visible; (slope approx. 1 in 150). Some disruption to drain, plumbing or heating pipes may occur. Damage to internal walls is more widespread with some crack filling or replastering of partitions necessary. Doors may have to be refitted. Inspection reveals some voids below slab with poor or loosely compacted fill.
4	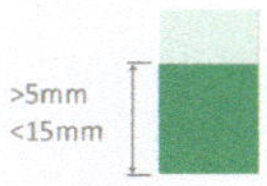	5–15mm (depends on number of cracks)	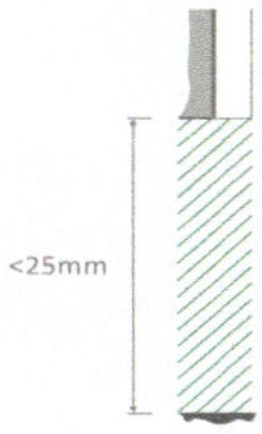	19–25mm	Large, localised gaps below skirting boards; possibly some cracks in floor slab with sharp fall to edge of slab; (slope approx. 1 in 100 or more). Inspection reveals voids exceeding 50mm below slab and/or poor or loose fill likely to settle further. Local breaking-out, part refilling and relaying of floor slab or grouting of fill may be necessary; damage to internal partitions may require replacement of some bricks/blocks or relining of stud partitions.
5	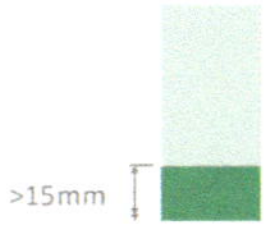	Greater than 15mm (depends on number of cracks)		Greater than 25mm	Either very large overall floor settlement with large movement of walls and damage at junctions extending up into 1st floor area, with possible damage to exterior walls, or large differential settlements across floor slab. Voids exceeding 75mm below slab and/or very poor or very loose fill likely to settle further. Risk of instability. Most or all of floor slab requires breaking-out and relaying or grouting of fill; internal partitions need replacement.

Note: *The 0–5 classification attempts to quantify the assessment of floor slab settlement damage in a way similar to that for superstructure damage, given in Table 6.1. It has not yet been used extensively to determine its applicability. It should be noted that the categorisation may be qualified by the possibility of progression to a higher category; this should arise only when examination has revealed the presence of voids or areas of loosely compacted fill (or degradable material) beneath the floor slab such that more settlement can be expected.*

**'Gap' refers to the void created — usually between the skirting and finished floor — caused by settlement after making appropriate allowance for discrepancy in building, shrinkage, normal bedding-in etc.*

There is general consensus that cracks of less than 1mm are trivial in structural terms, provided they are not varying in size seasonally by more than 1mm, or are not part of an array of cracks of greater (and increasing) total size. Indeed, this may also be applicable to cracks up to 5mm, after monitoring has shown that the crack is not progressing. However, it is often difficult to persuade Owners that 1–5mm cracks are of no structural importance provided they are stable and the cause has been identified and dealt with. Even the slightest hairline crack in decoration can be viewed with alarm, primarily because of the possibility of the cause being subsidence and the potential difficulty in selling the property.

The actual location of a crack often influences its acceptability. A hairline crack in a lounge ceiling can cause greater concern to a Homeowner, compared with the same crack in the ceiling of the kitchen or in the wall of a secondary bedroom. Yet relatively large cracks are often tolerated without any comment in a factory or commercial premises.

It is generally accepted by Experts that it is not the size of the crack/s that is important, but whether it is likely to increase in size if no action is taken. If a crack has reached or is likely to approach 5mm in width at its greatest dimension, then it is essential to be satisfied that:

- the stability and integrity of the property has not been affected by the damage
- the damage is not likely to increase further

Provided these two important factors are satisfied, the size of the crack or its effect on the appearance of the property is primarily cosmetic and only justifies localised treatment. In the case of an array of cracks, these factors need to be satisfied if the widths of individual cracks total 15mm or more.

Cracks of 1–5mm are sometimes described as being at the damage threshold or serviceability limit — beyond which the Owner, Lender, Warranty Provider and Insurer may consider that further technical investigation or assurance is required. Smaller cracks, provided they are not increasing in size may be considered as acceptable.

6.4.2.3 The timing of the damage
Note that there is an economical life for a building, depending on the quality and type of initial construction and its subsequent maintenance. This could be as little as 30–40 years for some types of construction, but usually averages about 100 years. Localised deterioration and damage may occur for many reasons during the course of that period. Cracking, whether caused by subsidence or any other factor, is one form of deterioration and must be assessed objectively, as with any risk.

6.4.2.4 Proximity of property to trees and vegetation
Note: In this guidance 'vegetation' means large woody shrubs and hedges.

Approximately 70% of all subsidence problems can be attributed to the behavior of clay soils, and approximately 60% of clay-related subsidence is exacerbated by the presence of a tree, multiple trees, other vegetation or a combination of trees and vegetation. Therefore, almost half of all structural subsidence problems involve trees, and in a drought year this can increase to 70% of all claims.

If the Expert knows definitively from local knowledge that the property is not founded on clay soils, the proximity of the trees/vegetation is unlikely to be a significant contributing factor to the damage. However, if it is suspected that clay is present, there is a strong likelihood that it is contributing to the problem.

The likelihood of damage arising from moisture extraction because of tree roots growing close to, or beneath a property, is very complex and depends on a number of factors including:

- The species, crown size, and condition of the tree/s or vegetation
- Their proximity to the property and depth of the property's foundations
- The climatic conditions
- The nature of the substratum including the shrink and swelling potential of the subsoil

There can be considerable variation within each of these factors, and their interaction becomes even more variable and unpredictable. As a result, very occasionally roots can cause building movement leading to damage at very considerable distances from a tree.

It is necessary to identify if a tree or trees are a contributory factor. A starting point is to identify the tree species and then to refer to Table 8.1 in this guidance to assess proximity.

There is also a need to assess the size of a tree; various methods exist, ranging from crown size assessment to mature height measurement. However, it must be remembered that if the tree has not reached its maximum mature size it will do at some stage. The advice of an arboriculturist will often be beneficial in this situation.

Trial pits at the damaged structure will show if roots have reached the structure; if roots are present they must be attributed to ensure accurate remedial action involving trees. Vegetation close to walls (Figure 6.7) cannot be ruled out as a contributory factor.

Figure 6.7: Large shrubs have the ability to cause damage when grown in close proximity to buildings

Further investigations (Chapter 7) should always be undertaken to support any consideration of the influence of any trees prior to any remedial action to the trees being undertaken.

This risk of damage will increase the closer a large tree is to the building. Buildings with shallow foundations will always be more vulnerable. However, a large proportion of trees including those on clay soils can co-exist in close proximity to a building harmoniously. Most trees, even large ones, but particularly low water demand trees, can co-exist in relatively close proximity to a building in granular or non-cohesive soils without ever causing damage due to these soils' non-shrinkable nature.

The depth of the foundations is unlikely to be established during a purely visual inspection. However, if cracks are present and if they appear to indicate subsidence, the likely position of maximum subsidence should be established. The vicinity of the property should also be inspected to see whether trees are present nearby and whether they are of a species likely to cause problems, or whether there is a tree stump of recently felled tree nearby. The distances of any trees or stump from the property are important criteria but as Chapter 8 confirms, the latest research shows that the crown volume of the tree is of greater consideration than the distance of a tree when assessing the likely moisture uptake demands.

Trees which are very close to a property (i.e., within 1.2m)[41] may cause local damage as a result of the mechanical action of tree growth pressing against the structure, whether or not the substratum is shrinkable clay (Figure 6.8). Greater distances are applicable when lightweight structures (porches, conservatories, etc.) are involved as they often have shallower foundations.

6.4.2.5 Proximity of the property to sloping ground or geo-features

Buildings situated on, or at the crest of, sloping ground can be sensitive to ground movement because the ground may have a natural tendency to move from high level to low level as part of the normal erosion process. The likelihood of movement occurring will depend on many complex geotechnical factors beyond the scope of this guidance. In simple terms the steeper the slop the more sensitive to movement. Soil slopes are more sensitive to movement than solid material such as rock, but even rock must not be dismissed as it may have internal fractures and fissures that could permit ground movement.

Figure 6.8a and b: Cracking due to tree root-related subsidence damage

Substrata consisting of clay interspersed with thin layers of sand that permit water to percolate through the ground are particularly sensitive to ground movement. Such movement (landslip) occurring in the vicinity of a property can result in subsidence. Indications of an unstable slope include:

- Displacement and/or out-of-plumb fence posts, trees, etc.
- Fissures on the surface of the ground
- Local 'slumped' areas of ground surface

Although buildings on, or at the crest of, a slope are more likely to be at risk from ground movement of this sort, properties close to the foot of sloping ground may also be sensitive to the results of the landslip if the ground should move dramatically towards the property.

Slopes created by recent ground excavation (such as excavation for adjacent foundations or basement construction) or filling operations should also be viewed with great caution. Excavations undertaken at the foot of an embankment can cause instability, as can load material placed on a slope, particularly towards the crest. If a slope or embankment already has a drainage system within it, the blockage of the system or interference with it can alter the groundwater conditions, which might give rise to instability and landslip.

Granular materials, such as sand and gravels, can remain stable for long periods provided they are not physically washed away by surface water. Slopes of up to 30° for such materials are normally not a problem. However, in the case of a clay substratum, much shallower slopes can be sensitive to ground movement and it is advisable to be wary of ground which has a slope greater than approx. 8°. Even shallower slopes may be unstable if the clay is fissured.

Reference should be made to a professional with suitable geotechnical experience for proposing advice on this subject.

Although the initial appraisal will not be capable of identifying whether the ground is likely to move or not, it will identify the presence of sloping ground and particularly whether a property shows any historical signs of downhill movement.

6.4.2.6 The likelihood of leaking drains

If a property displays subsidence-type cracks the likely location of the subsidence should be identified from the crack pattern, and a superficial examination made of the immediate vicinity to see whether drains are likely to be present (Figures 6.9 and 6.10).

Leaking drains can cause subsidence in two ways:

1. Water can leach out into the ground and soften the substratum beneath the foundations, rendering the ground incapable of supporting the weight of the building. This mechanism can happen irrespective of the depth of the drains relative to the foundations.
2. Alternatively, where drains are located close to and below the foundations, material can be washed out into the drains from beneath the foundations, creating voids.

Figure 6.9: Subsidence at corner of property due to leaking drains

Figure 6.10: Effect of broken or leaking drains

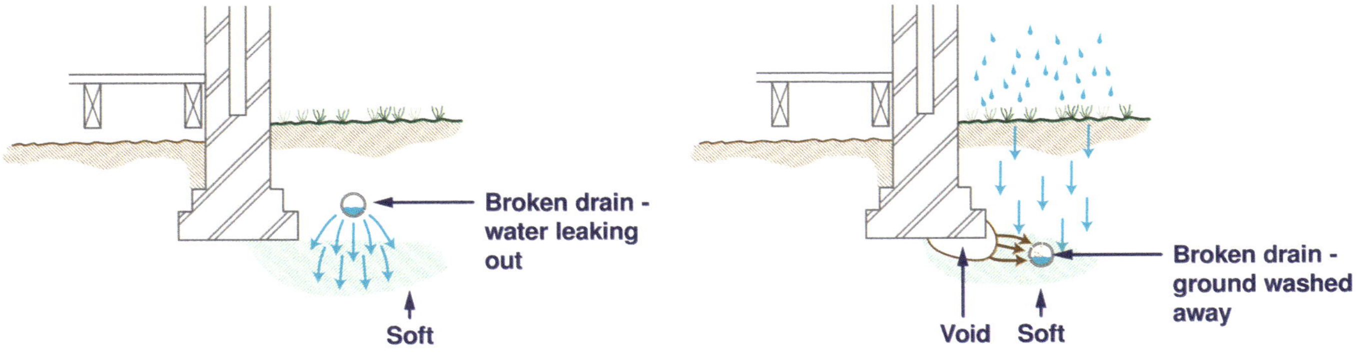

Properties built before c.1970 in the UK will invariably have drains constructed with rigid joints. This type of drain is particularly susceptible to developing leaks. More recently, drains have incorporated flexible joints which are less likely to leak, but broken pipes or bad drain laying can be a common problem with either rigid or flexibly-jointed drains. Consequently, it should not be assumed that defective drains are unlikely to be a cause of subsidence just because a property is relatively new. Tree roots can cause direct damage to drains when roots enter the drainage pipework via hairline cracks or inadequate joints in the pipework.

As the initial appraisal does not involve the excavation of trial pits, any assessment as to whether leaking drains are likely to be the cause of the problem will be somewhat subjective but should be reported if suspected — particularly where drainage pipework or water discharge is located in proximity to the maximum point of downward movement to the property.

6.4.2.7 Likelihood of the property being on subsidence-sensitive soils (local knowledge)

It may be known to those making the initial appraisal that a clay substratum or other ground conditions exist in a particular area in which problems have occurred in the past. Large-scale geological maps will assist in establishing the subsoil stratum. Note, that the majority of subsidence problems affecting low-rise buildings, however caused, generally occur within 4–5m of ground level. Geological survey maps do not accurately show shallow deposits, hence individual investigation is invariably required.

Knowledge of the substratum is particularly useful if tree root penetration beneath or in the vicinity of foundations is thought to be a problem. As well as knowing the existence of shrinkable clay in any area, knowledge of the existence of other factors such as the presence of peat, current or past mining operations, quarrying, etc., all constitute valuable local information which can considerably assist the appraiser. A non-specialist Expert having such local knowledge, e.g., a Valuation Surveyor, Architect, etc., may be able to put the problem in context.

Local knowledge should be considered with a degree of caution however, as it can be easy to over-generalise e.g., although properties in a particular area suffer from subsidence due to the known presence of potentially troublesome underlying stratum, e.g., peat, those nearby may not.

6.5 Subsidence crack patterns

6.5.1 Identification

Subsidence causes foundation movement, but the evidence of such movement is normally noticed first in the form of cracks appearing in the superstructure.

The vast majority of domestic properties (and older commercial and industrial buildings) in the UK, have been built using masonry, i.e., stonework, brickwork or blockwork. Masonry is a material that is good at resisting compressive forces such as direct loading, but weak at resisting tensile forces.

Modern masonry structures suffering from movement invariably develop cracks, but different types of movement tend to give different crack patterns. Thermal and shrinkage movements do not generate the same types of crack patterns as subsidence (Figure 6.11).

Figure 6.11: Crack in stud wall due to sagging or shrinkage of supporting timber joist (non-subsidence)

Buildings which are framed, e.g., in steelwork, or clad with timber boarding, metal sheeting, or which have tile or slate cladding do not display evidence of subsidence as readily as buildings constructed from masonry. Fortunately in the UK, low-rise domestic properties with cladding on the external elevations tend to have internal masonry, even if it is plastered or rendered. As such, it is still usually possible to identify subsidence movement within buildings clad externally with sheeting, by examining for cracks on internal wall surfaces. Care should be taken to check that the walls themselves are not covered in cladding, e.g., dry lined, etc.

In practice, subsidence cracks are best identified by examining external masonry elevations because this allows large areas of a wall to be viewed at once, particularly on two- or three-storey buildings (Figure 6.12). Only relatively small areas of wall can be examined in each room in turn inside such a building and it is possible that internal decoration will, in any event, have been repaired — hiding all but the most recent cracks.

Figure 6.12: Crack in external wall (subsidence)

External cracks are more likely to be easily identified, even when the masonry has been repointed (Figures 6.13 and 6.14). Standing back from an external elevation may make it possible to note cracks which have been repointed, along with any other distortions in the coursing of the masonry which would indicate past movement. External examination also allows complete crack patterns to be easily identified.

Figure 6.15 shows a typical array of cracks in relation to subsidence of the shallow founded section of the property.

From the early/mid 1800s until c.1960, lime mortar was used in the construction of masonry walls. Cement mortar used today is relatively brittle compared with lime mortar and tends to show distortions and structural movement by forming conspicuous cracks.

Cracks in cement mortar can usually be easily identified even when as slender as 0.5mm in width. By contrast, lime mortar is somewhat more plastic and so can accommodate slight distortions of the masonry without showing small

Figure 6.13: Repointed cracks in render

Figure 6.14: Individually repointed cracks can be conspicuous

Figure 6.15: Array of cracking damage

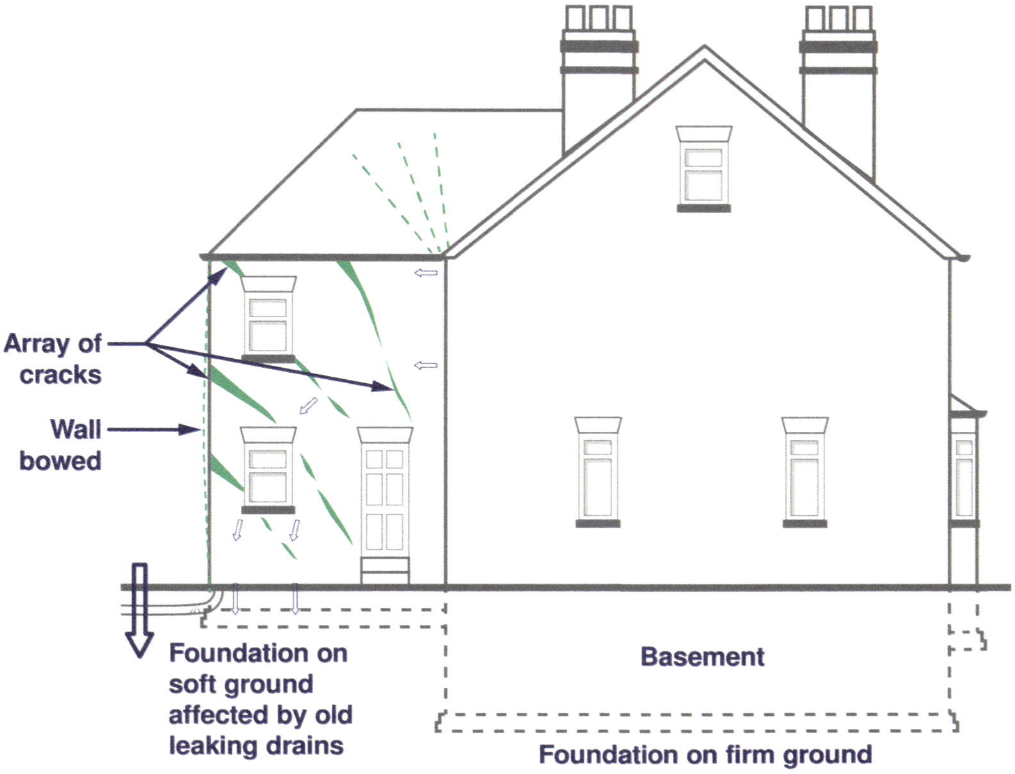

cracks. However, lime mortar joints in masonry tend to weather more rapidly than cement mortar joints and so need to be pointed or repointed approx. every 20 years. The pointing usually consists of the harder cement-based mortar, and therefore more recent subsidence movement tends to show up by cracking of the brittle pointing.

6.5.2 Characteristics

Subsidence cracks are generally identified by the following characteristics:

- They tend to be diagonal in direction
- They tend to be wider at one end (Fig. 6.15)
- They always start and are widest at a 'free edge', such as the corner of a house, the top of a wall or the end of a wall, or the edge of a door or window. The free edge can be at the foundation level in rare cases such as heave
- They tend to travel around openings
- They often extend below damp-proof course (DPC) level

Figure 6.16 illustrates a typical pattern of cracking where subsidence has occurred.

There are several identifiers that the Expert should look for:

- Cracks are always widest at the end which meets the 'free edge'
- Cracks may extend from the free edge to another free edge, or may disappear at an intermediate location within the area of the wall. It is, however, only the narrow end of the crack which will 'disappear' at any intermediate point in the wall (Fig. 6.16)
- A crack which exists on an external elevation will almost certainly be matched by a corresponding crack/s on the interior face the wall. Such internal cracking may have been made good in the past by filling and redecorating but it must, in the vast majority of cases, have existed at some time. A crack on one side of a wall does not always correspond exactly on the other side. It is more likely to correspond if the wall is of solid construction, but where a wall is of cavity construction the crack might be displaced by some distance, to another location on the other side of the wall. It is important to appreciate that a wall cannot crack on one side from subsidence without also cracking on the other

Figure 6.16: Typical pattern of cracking

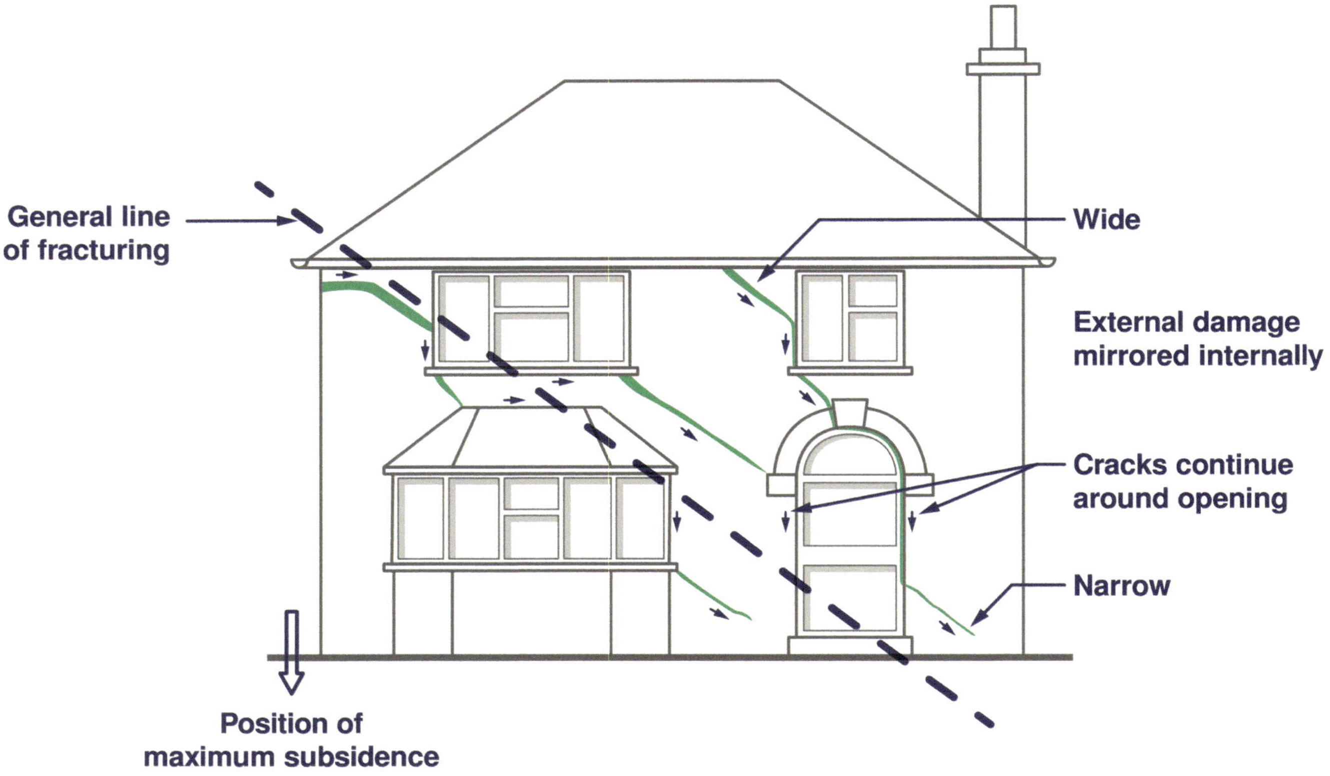

- Openings in a wall, such as doorways or windows, tend to attract subsidence cracks because they find it easier to travel around the opening than to fracture the adjacent masonry. Openings in a wall should be considered as part of the crack pattern, e.g., a crack arriving at the bottom right hand side of a ground floor window may continue at the top left hand side of a window, but will be a little wider. In each case the window has merely been incorporated into the crack

Sometimes a wall will display a single large crack being, for example, 1mm wide at DPC level near the ground and extending up to a higher-level edge where it may be 10–15mm wide. Alternatively, the same amount of distortion may cause a series of cracks to fan out from near the same position at ground level. Such a series is normally referred to as an 'array'. Where two or more cracks result from the same amount of foundation subsidence they will be narrower than a single crack caused by the same amount and pattern of subsidence.

It is important when considering cracks that the total number of cracks making up an array are identified and the sum of their maximum widths at the free edge or edges should be added together to get the equivalent 'single crack' width.

Cracks which arrive at a free edge will continue around the edge and either up or down the adjoining wall, depending on the type of subsidence.

It is important to appreciate that a property suffering from subsidence may be experiencing the effect at more than one location. This can result in a complexity of cracking which is sometimes difficult to interpret (Figure 6.17).

The manner in which the building is moving will often become clearer if the person undertaking the initial appraisal sketches the external elevations of the property, and, if possible, the internal elevations marking the cracks on the sketch, noting particularly the locations where cracks meet free edges, as well as points of maximum and minimum crack widths. This should enable the location of the movement to be identified with some accuracy, but in some circumstances may require level monitoring (Section 7.3.4).

Where heave cracks have occurred (normally as a result of tree removal permitting a shrinkable clay substratum to swell again) they tend to have a similar pattern to subsidence cracks but 'in reverse'. However, cracks resulting from heave can display complex patterns. Much depends on which part of the building is heaving relative to another.

Figure 6.17: Typical damage resulting from two separate causes

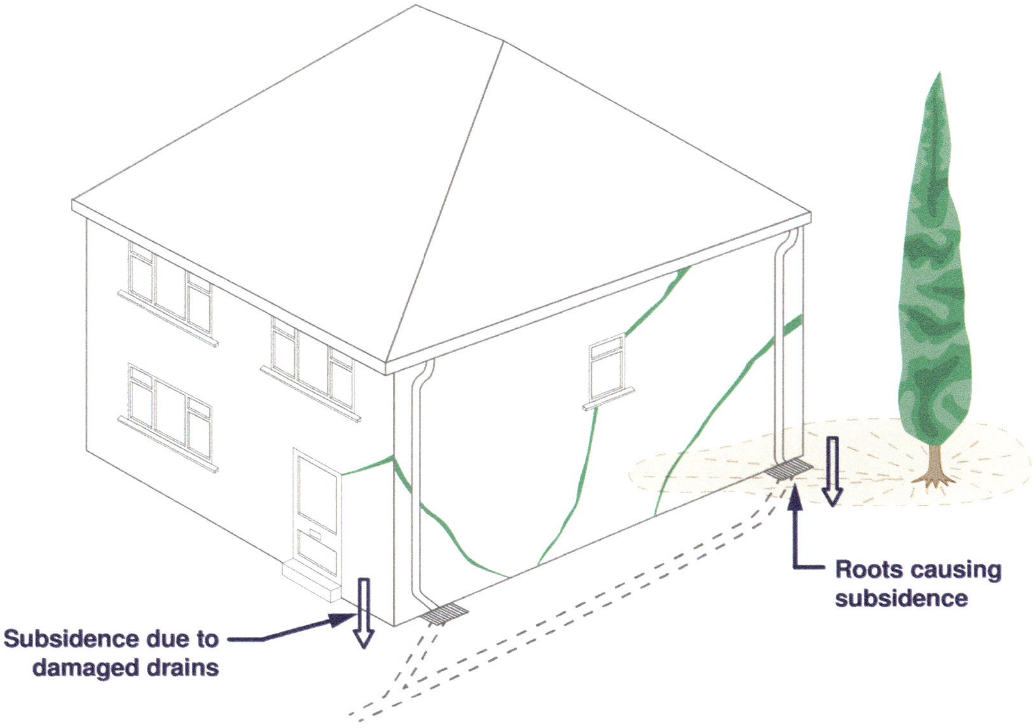

Care should be taken when interpreting heave movement to confirm which parts of the structure are moving and which are stable. Heave cracks may not display a diagonal tendency to the same extent as subsidence cracks and can frequently be almost vertical but, like subsidence cracks, they are always tapered, with one end wider than the other. It is also possible that the tree may have caused initial subsidence cracking which might have been repointed, confusing the crack patterns.

Once the location/s of the movement have been identified, there are two simple procedures which often help to confirm the diagnosis without having to excavate trial pits or cut into the fabric of the property:

1. If it is possible to sight along the horizontal bed joints of the masonry on the relevant external elevations, it should be possible to see if the coursing dips down from the point where the crack or the array has been identified. In Figure 6.18 the horizontal bedjoint of the houses with the lower gutter line appear to drop and not level with those on the right-hand side of the vertical joint on the elevation. This procedure tends only to identify a discontinuity in the wall and it is important that the section of wall which is actually subsiding is identified. It will always be the part under the diagonal crack – which will be the part of the wall closest to the free edge.
2. A possibly less reliable method, is to place a spirit level on internal floors in the vicinity of the movement to see whether the floors dip down towards the location where it is thought that the subsidence has occurred (or rise up in possible heave conditions). When checking floor levels it should be appreciated that timber floors in particular can sag and distort for various reasons not at all associated with subsidence, and in the case of longstanding subsidence, floors may have been relevelled many years before, so that any distortion which did exist has been removed. Concrete floors can suffer from their own particular problems, e.g., sulphate attack, creep/sag of prestressed units, settlement of fill underneath the oversite slab, etc.

Plumb line and level surveys would normally constitute part of the further investigation but on some occasions it may be appropriate to perform them at this stage. Detailed information is given in Chapter 7.

6.5.3 Decorations

Knowing the age of internal decorations can be very useful when diagnosing the importance of subsidence damage. Cracks in a newly decorated property would suggest recent movement and may give rise to greater concern than minor cracking in much older decoration. In the latter case, the movement at least indicates the total degree of

Figure 6.18: Brickwork coursing has lowered on left-hand side

movement which has occurred since the decoration took place, but does not necessarily indicate whether movement is recent or long standing or, for that matter, whether it has been gradually occurring over a number of years since the previous decoration or is a more recent phenomenon.

In properties where heating has been provided by open fires, decoration can look much older than is actually the case. Recent installation of central heating can cause shrinkage cracking due to the drier atmosphere.

Where a property has suffered previous subsidence, but where internal decoration or external pointing is clearly at least five years old and no fresh movement is noticeable, the subsidence is likely to have stabilised.

Hairline cracking, i.e., less than 0.5mm wide, at positions where previous cracking has occurred should not automatically be assumed to be of structural importance or to be indicative of continuing movement. When a wall has cracked, for whatever reason — not necessarily subsidence — it is fractured and although is probably quite stable, the mere repointing or filling and redecorating over a crack will not mechanically rejoin the masonry either side of the crack.

Hairline cracking is very rarely of any importance from a structural point of view. Seasonal variations between hot and cold and wet and dry periods can cause very slight flexing of the structural fabric which can manifest itself as hairline cracking at the point where the old crack previously existed. In such instances a crack up to 2mm wide should not normally give cause for concern providing it is of a seasonal nature and not progressively increasing in size.

6.6 Report

Following the initial appraisal, a report is required to record the findings and provide an assessment of the initial visit. This should ideally include the following:

- The reason for the report and who instructed the appraisal
- The limit of the report, e.g., 'preliminary based on external inspection only'
- A brief description of the defects, ideally with sketches/photographs/video
- The likely cause of the defects including comment on any other possible causes
- A statement on the severity of the defects
- Possible prognosis — with reasoning, and including any concerns
- Any further investigation required
- Recommended course of action
- Any likely repairs required with estimated costs, where causes can confidently be established

Often the report is required to confirm that the movement is 'non progressive' i.e., has stopped or is very unlikely to continue/increase. A response to this should be carefully framed with supporting evidence.

The Expert should remember that the report is their opinion based on the evidence gathered and considered. The Expert is not expected to be infallible but should exercise all such reasonable skill, care and diligence as is to be expected of a properly qualified and competent Expert experienced in this type of work. Great care should be taken in the preparation of the report not to overlook important issues or to give assurances that cannot be supported.

The report will be read by those who are not necessarily experts in building structures and should be written with clarity in mind, with straightforward, unequivocal conclusions. BRE Digest 251[6] gives very good advice on the reporting of damage and assessment of severe damage. Many organisations now have their reports in standardised formats to ensure a level of consistency and reduce risk of omitting vital information required in decision-making.

It should be remembered that the report may be required for future purposes including for example as part of any recovery action against a third party. In this instance, it is imperative that a full set of photographs and description of the damage (including site notes) are prepared and recorded, following the initial visit.

7 Further investigations

7.1 Objectives

The initial appraisal should have identified whether damage has been, or may in future be, caused by foundation movement, and identified possible causes for that movement. Further site investigations may be necessary to:

- Confirm or determine the exact cause of damage
- Determine the extent of subsidence/heave within the building
- Determine whether it is likely to continue:
  - o Progressively (in the same direction)
  - o Seasonally (annual cycle of subsidence and recovery)
- Determine whether any continuing movement will exacerbate the damage
- (if unacceptable movements are likely to continue) Determine the most effective remedial solution to stabilise the property and repair any damage
- Confirm the adequacy of remedial action

7.2 Scope

Different investigations may be necessary for different causes and for the various remedial options. For this reason, only some of the investigations described in this chapter may be appropriate for a particular case. The initial appraisal may have identified which are required. If it hasn't, they should be identified at the outset, and then modified as necessary as further information becomes available.

As investigations proceed and the cause is confirmed, the scope of investigations may need to be modified. For instance:

- If trees are identified as a potential cause, before their control can be considered as a potential remedy, investigations need to establish whether a persistent moisture deficit is present (Section 5.2.3.3)
- Where a stabilisation scheme is identified as the probable remedial measure, investigations should include:
  - o the foundations, and ground and water conditions in areas of subsidence/heave
  - o a suitable bearing stratum, including ground and water conditions within which the stabilisation system will be controlled

The cost-effectiveness of investigations should be continually appraised to ensure that all necessary information is obtained, but without waste of resources on unnecessary work. A fixed pattern of investigation for all cases of subsidence is unlikely to be appropriate. In fact, it has become more common to avoid investigations completely, if causation can be confidentially identified from a visual appraisal. However, parties need to be aware that for complete comfort over diagnosis, thorough investigations into the cause are recommended.

For investigation works, numerous field and laboratory specialists operate in the subsidence marketplace, providing trained and well-equipped crews to assist and provide factual reports for the subsidence professional to act on.

7.3 Assessment of foundation/superstructure

This element of the investigation is effectively a recording of the degree of damage and distortion, supplemented by the monitoring of any ongoing changes.

7.3.1 Full crack record

The initial appraisal is likely to have been based on analysis of the crack pattern (Section 6.5). As confirmation of this, a full written record of the damage should be made, usually in the form of a sketch or detailed plan of each elevation, and an 'exploded' plan of each room showing walls and ceilings. The location and dimensions of every significant crack should be recorded, noting the position where crack widths are measured.

Examination of damage should include a record of previous damage, including assessment of whether historic damage is consistent with current problems, or indicative of some other cause.

Refer to the BRE's *Simple measuring and monitoring of movement in low-rise buildings* (Part 1)[46] for further information.

7.3.2 Distortion survey

Damage is unlikely to have occurred unless there has been significant differential movement between different parts of the property. If it is assumed that brickwork was constructed level in the first instance, any departure from level will indicate differential movement. Variations in level can be detected by levelling along a convenient brick course, preferably as close to the damp-proof course (DPC) as possible. Measurements can be taken with a water level, but as a precision optic level (including telescopic levelling staff) will be required for level monitoring (Section 7.3.4), it is usually simplest to use the same equipment. Readings should be taken at regular intervals around the building, perhaps every 2m in areas of little movement, but at closer (1m) intervals if significant differences are apparent. Readings to either side of door or window openings are desirable. Results can be calculated either relative to an arbitrary point, for instance adjacent to the front door, or relative to the highest or lowest point on the brick course. They can be shown either as plus (+) or minus (-) values on a simple plan, or by vertical bars on an isometric plan (Figures 7.1 and 7.2). Measurements can also be taken internally on a floor slab, either just around the periphery of each room or with a lattice of measurements over the whole area.

Figure 7.1: Typical results for a level distortion survey, with levels presented in mm relative to an arbitrary zero datum adjacent to front door

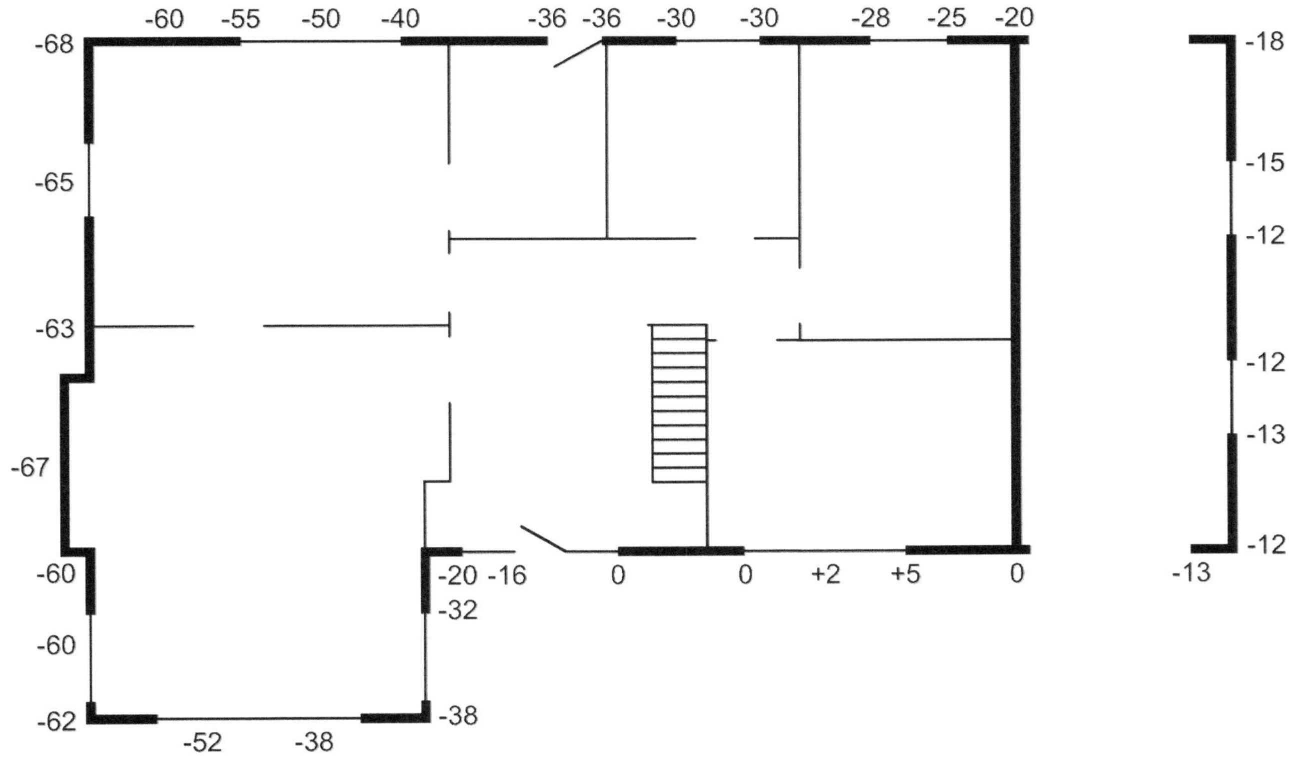

The level survey can be supplemented by a verticality survey of external walls. As this needs to measure the full height, a builder's level is inadequate; a plumb bob or theodolite with an offset staff should be used, with measurements taken at approx. 1m intervals. Sets of measurements should be taken at each corner, and midway along each wall. Results are usually best presented by horizontal bars on an isometric drawing of the elevations. Manual execution of verticality surveys is becoming less common, with technicians now able to set up total station devices to establish distortion and movement far more rapidly and with more accuracy.

Both level and verticality surveys are based on the assumption that walls were originally constructed with horizontal courses and vertical perpends. This is normally a reliable assumption. Note however, that any recorded distortion could

Figure 7.2: Level distortion survey presented on an isometric plan, using values from Fig. 7.1

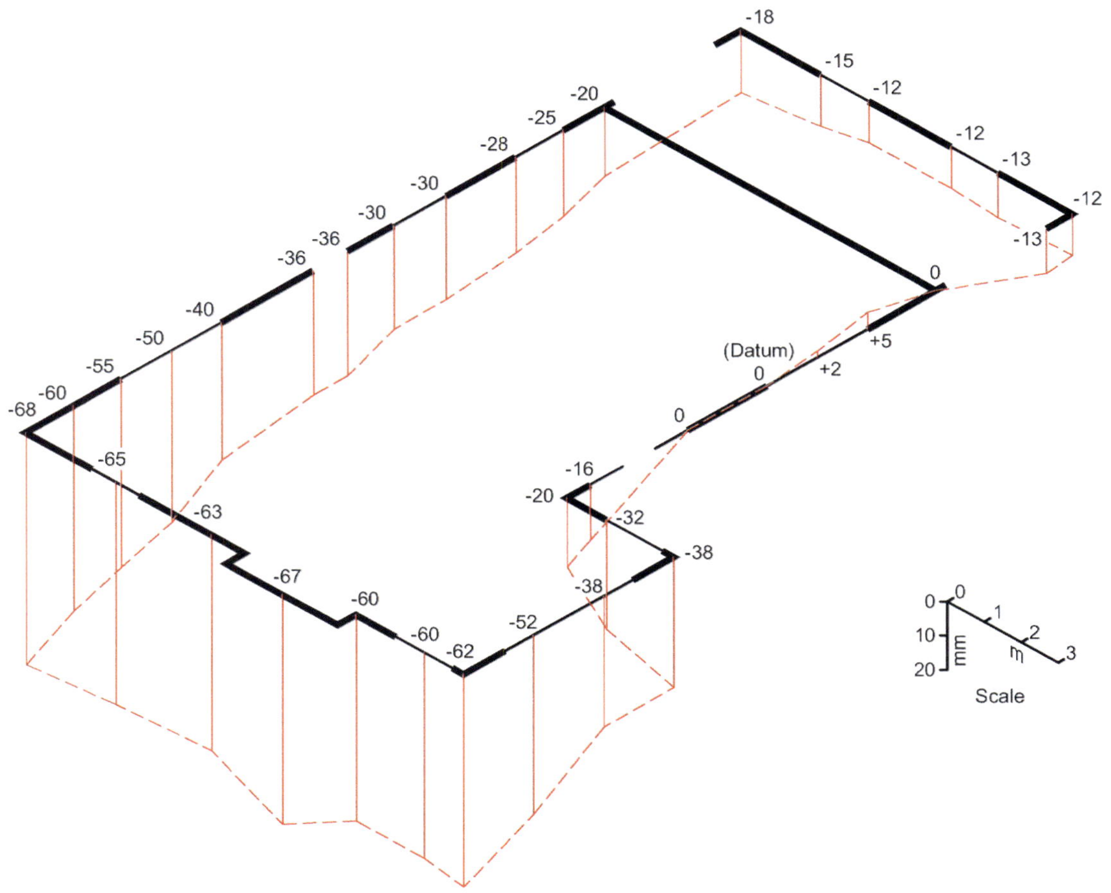

have occurred at any time in the life of the building, or be a legacy of the original construction; and may not indicate recent movements. Results from distortion surveys can be difficult to interpret and require experience and skill. The whole pattern of movement should be considered rather than individual pieces of evidence.

Refer to the BRE's *Simple measuring and monitoring of movement in low-rise buildings* (Part 2)[47] for further information.

7.3.3 Crack monitoring

Crack monitoring[46] can be regarded as the standard form of monitoring because sufficient accuracy can be achieved without relying heavily on skilled operatives and specialist equipment. As well as its use in diagnosis, crack monitoring has value in confirming that remedial action to stabilise foundations has also been effective in stopping further crack movement. When used for this purpose it may be possible to delay the start of crack monitoring until after remedial action has been taken, but as information on the behaviour of cracks may also be useful for diagnosis, an early start to crack monitoring is usually desirable. Crack monitoring is particularly necessary if the damage relates to superstructure problems rather than subsidence. Superstructure repairs and redecoration (Section 9.6) should not be initiated until crack monitoring has demonstrated that significant crack movements have ceased.

If the results of the exercise are to be useful, the readings must be accurate, i.e., capable of detecting movements of less than 0.1mm. The only reliable way to achieve this is with the aid of instruments measuring from fixed markers (Figure 7.3a). Glass 'tell-tales' (Figure 7.3b) can be inaccurate and vulnerable to vandalism, but specialist calibrated tell-tales are effective. Callipers, either with an electronic digital display or with a vernier scale, provide sufficient accuracy (Figure 7.3c). The most suitable markers are the small stainless steel studs intended for use with digital callipers, glued to brickwork, plaster or other surface with a fast-drying epoxy resin.

Monitoring should be for a specific purpose and for a specific period, typically ranging from 6–12 months but exercises can be extended depending on circumstance.

Figure 7.3a: Measurement using a crack measuring gauge (basic)

Figure 7.3b: Measurement using glass tell-tales (not recommended)

Figure 7.3c: Stainless Steel studs with digital callipers

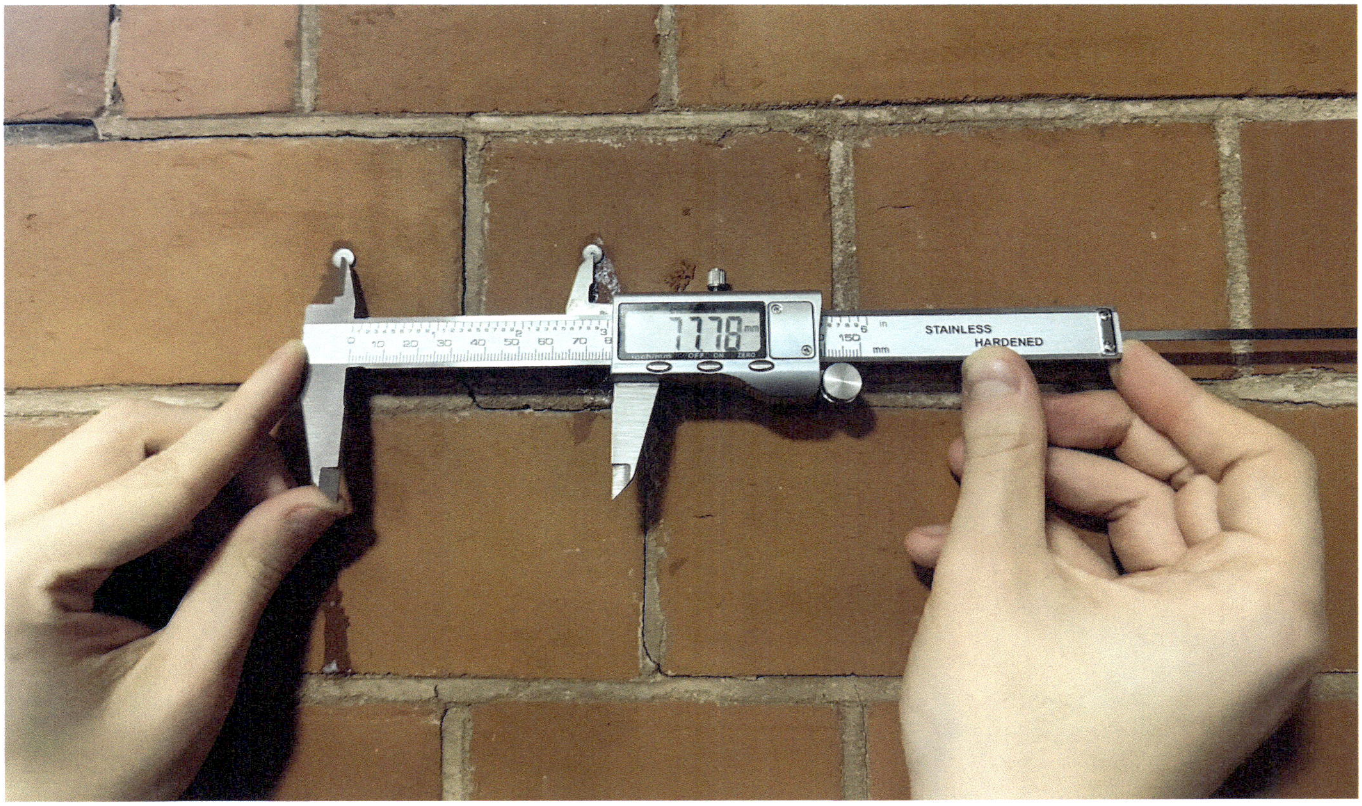

7.3.4 Level monitoring

Some Experts argue that level monitoring is preferable to crack monitoring because subsidence damage is caused by movement of the foundations. It is therefore logical that investigations should include direct measurement of these movements. This type of monitoring will show exactly which parts of the building are moving and by how much, and so should be an important element of most investigations. Different causes of subsidence produce different patterns of movement, e.g.:

- Trees or other vegetation – seasonal, subsiding in summer and recovering in winter. The amount of subsidence and recovery may not be equal
- Heave as a result of tree removal – progressive upwards, at a diminishing rate
- Settlement, erosion, mining subsidence and slope instability – progressive downwards, possibly accelerating in wet weather

In addition, if trees or shrubs are involved, and remedial action requires control of this vegetation (Section 8.3), level monitoring will be the most effective method of demonstrating the adequacy of the remedial action. If the results of level monitoring are to be valuable they must be accurate, aiming to achieve a closing error around a typical building of less than 1mm. To achieve this, readings should be taken to the nearest 0.1mm, and should use suitable equipment, such as:

- Markers – which are firm and robust, and provide an accurately defined location from which to measure. A brass or stainless screw of adequate length, mounted into a plastic plug close to ground level will usually prove adequate
- A telescopic levelling staff – which should mount accurately on the marker and can be held vertically against the wall, consistently, to exactly the same position (Figure 7.4). It should be marked in such a way as to allow readings or interpolation to the nearest 0.1mm
- A precision level – that is regularly maintained and calibrated

Figure 7.4: Level survey in progress

Accuracy is improved if exactly the same procedure is followed for each set of readings, i.e., using the same operator and equipment, and setting up in the same position.

Markers should be attached at regular and appropriate intervals around the building e.g., at each corner, plus 2–3 intervening positions along each wall, at the corners of any projecting elements and at the junction with any extensions. Markers on internal loadbearing walls, which can be observed through a door or open window, are often useful.

In most situations it will be found that parts of the building are showing little or no differential movement, and one of the markers in this part of the building can be selected as the datum. However, if the whole building is moving, and particularly if several contributing factors are active at the same time, an independent datum will be required. In some situations it is sufficient to use an accurately identifiable part of a manhole surround, particularly if this is deeply founded. A manhole cover, which may be moved, is not reliable. For greater reliability, a deep datum may be required; this should be located in an area where the ground is unlikely to be subject to movement (for instance from tree roots or slope instability) and founded at a depth of at least 3m.

The results of level monitoring are usually best presented graphically, plotting the movements against time for each location.

Ideally, level monitoring should be initiated as soon as possible. The frequency of subsequent readings will depend on individual circumstances, but intervals of 6–8 weeks are usual. As results become available, preliminary conclusions can be made, and refined as further readings are taken. If vegetation control is adopted, it will usually be desirable to continue the monitoring to demonstrate whether the action taken has been effective (Section 8.5). Readings each spring and autumn will usually be sufficient for this purpose. If markers are inconspicuous, they can be left on a building indefinitely.

Refer to *Monitoring building and ground movement by precise levelling*[48] for further information.

7.3.5 Visual monitoring

In some situations, where a party wishes to repair their property without the benefit of a formal monitoring exercise, visual monitoring may be an option. In such circumstances, damage is repaired immediately after mitigation works to arrest further movement, with the renewed internal and external surfaces acting as a datum against which future movement can be gauged. If no damage is sustained, the immediate post-mitigation repair can be considered as having been a worthwhile risk, but if movement recurs, all parties need to appreciate that the funds invested in the speedy repairs will have been wasted and further work and expense will now be incurred.

7.3.6 Remote monitoring

Over the last 10–15 years technology has been developed by specialists to precisely monitor vibration and movement predominantly connected to major civil engineering and infrastructure projects. This involves accurately located sensors about a building with remote readings taken to establish activity, and data sent to a website host via mobile phone network. Use of this equipment is gradually appearing in the subsidence industry. It has the advantage of providing 'real time' movement, as activity is logged remotely and more continuously and temperatures can be recorded to assist diagnostics. However, it can prove relatively expensive compared to more traditional technician-lead solutions.

7.4 Soil investigations procedure

Physical investigations[49] can be relatively expensive. They are however, a very important element of the overall investigation and can yield important, accurate technical information. They should be properly targeted and carried out efficiently and economically.

The information obtained should be interpreted by an Engineer with appropriate experience of geotechnical matters. Various methods of physical soil examination are discussed in the following subsections.

7.4.1 Trial pits

Trial pits (Figure 7.5) are used to obtain details of the foundations and adjacent soil conditions. The number of pits required and their location/s will depend on the circumstances of the case and are usually best determined using the results of the distortion survey (Section 7.3.2). Depending on site conditions, they can be machine or hand dug. Using a mechanical excavator obviously saves time where access permits, but care should be taken to establish the position of any underground services beforehand and it is obviously safer to dig by hand. Even shallow excavations can collapse and the risks should be assessed; the requirement to take steps to prevent danger to any person is set out in CDM Regulation 22[15]. Care must be taken that they do not significantly undermine the foundations. A vertical trimmed face makes it easier to inspect and identify individual strata within the trial pit.

Figure 7.5: Typical trial pit

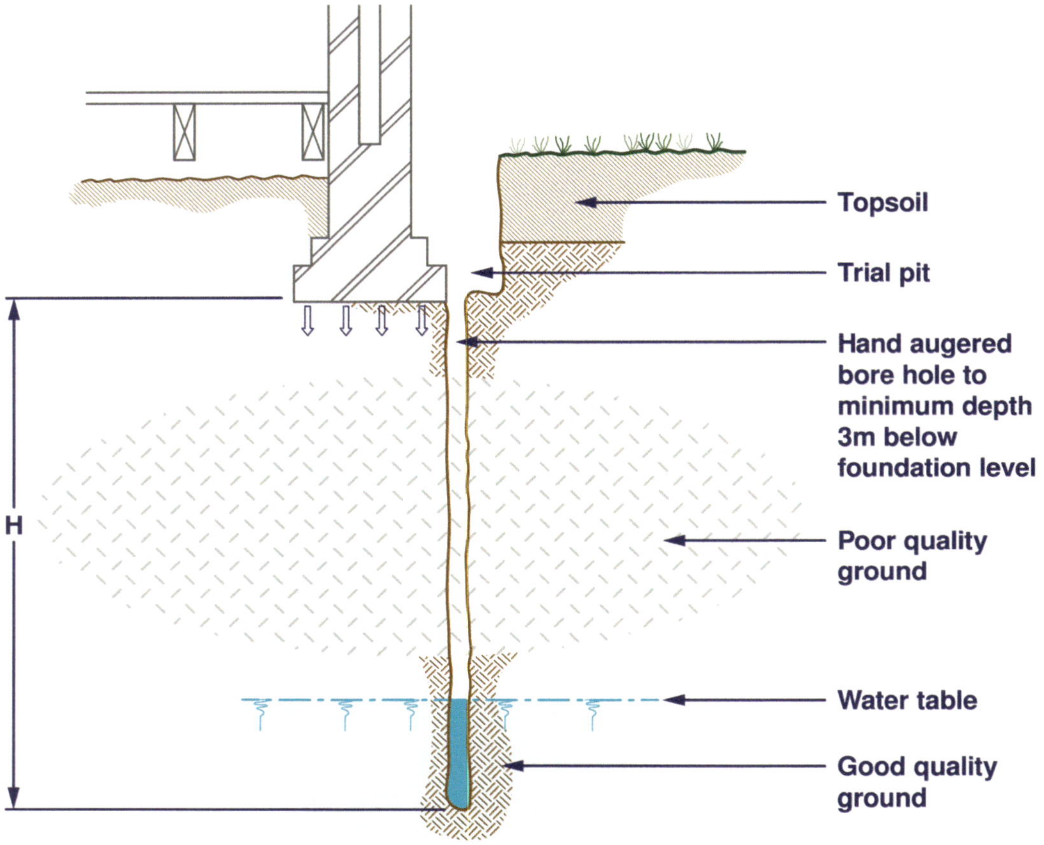

Information obtained from a trial pit should include:

- Dimensions and details of the construction of the foundations
- A good description of the ground beneath the footing
- A description of the ground alongside and above the foundations
- The presence of services, roots, etc.
- The presence of groundwater (levels found and final standing levels)

Simple *in situ* strength tests[50] can be carried out using, for example, a shear vane. Soil samples can be taken for testing to establish the shrinkage potential, etc.

Trial pits should be backfilled as soon as possible and not left exposed to weather for any period. Materials should be replaced in approximately their original sequence, and well compacted in layers.

Refer to DG 411[49], DG 381[51] and BS 5930[52] for further information.

7.4.2 Boring and hand auguring

Hand auger borings can be used as a means of extending trial pit depth, but they do not allow the soil to be inspected *in situ*. Hand auguring usually only permits the taking of disturbed samples (Figure 7.6) and only permits a subjective assessment of the consistency of soil strength at relatively shallow depths. The recovery of undisturbed samples, required for laboratory testing, requires the use of light cable percussion techniques (Figure 7.7) or demountable rigs (Figure 7.8) where site access is limited. These rigs are normally necessary where sampling is to be carried out at depths in excess of 3m. When using cable percussion rigs, *in situ* testing can be carried out, e.g., standard penetration tests and a distortion survey. Sufficient boreholes will be needed to determine the pattern of any variation in conditions across the site.

Figure 7.6: Hand augured borehole sample showing change in soil profile

If underpinning or piling is contemplated as the remedial measure, the depth of boreholes must be sufficient to establish a suitable bearing structure and strength testing of soils. Where tree root action is suspected, a comparison hole away from the potential influence of the tree may be needed, provided soil conditions across the site are likely to be sufficiently uniform to allow meaningful comparisons.

Refer to *Site investigations for low-rise buildings: direct investigations*[49] for further information.

7.4.3 Probing – penetration testing

The investigation of the depth of hard strata below weak soil is often carried out using the probing method. In its simplest form, the probe is a sharpened steel rod which is pushed or driven into the soil until it meets resistance which will be offered by a hard stratum. This method may be used effectively when establishing the extent of soft soils overlying harder ones.

Figure 7.7: Hand-held window sampler

Figure 7.8: Tracked rig

More sophisticated forms of probing have been developed. The two principal types are:

1. Dynamic probe (Figure 7.9), which is driven into the soil by means of a hammer blow. Dynamic probing can be used to interpolate soil conditions between borehole positions.
2. Static probe which is forced into the soil by static load.

Figure 7.9: Dynamic probe

With the static probe, a rod is pushed into the ground and the resistance at the tip is measured. Static probing is known by a number of terms including the 'cone penetrometer test'. This test is relatively quick to carry out and is obviously economical by comparison with boring, sampling and laboratory testing. Continuous soil profiles can be established by interpretation of the ratio of frictional and cone resistances.

All forms of probing can give unreliable results in soils containing occasional gravel, cobbles or boulders as these can easily be mistaken for bedrock. The cone penetration test was originally developed for use in alluvial soils but its use has been widened to include other materials such as chalk and clays. Depth of penetration is limited by the thrust available for pushing the penetrometer into the ground.

7.4.4 Soil examination

To achieve the maximum benefit from soil tests, it is important to record the following information when carrying out a physical soil examination:

- The type of material and its position, colour and odour
- The predominant particle size
- The state of compaction and the presence of vegetable matter
- The soil structure and texture, e.g., fissured or stratified
- Any major obstacles encountered
- The likely loadbearing capacity of the soil at various depths
- The moisture conditions
- The presence, depth and condition of any roots

7.4.5 Soil tests
Suitable samples should be taken at appropriate intervals from trial pits or boreholes for further laboratory testing. The number of samples, and the interval between samples, will depend on the type of tests. For moisture content determination, samples at close intervals (every 0.2m or 0.25m) are desirable, to allow accurate determination of the moisture content profile. The greater cost of other laboratory tests may dictate wider intervals, but it should be commensurate with the type of information required.

The following tests may be appropriate:

- Soil classification e.g., to determine liquid and plastic limit
- Moisture content determination, e.g., with depth
- Soil suction[53]
- Triaxial
- Sulphates and pH
- Organic content

7.4.6 Assessment of desiccation from soil testing
It is important to establish the extent and depth of any desiccation but this is not always easy. Various methods are available:

- Comparison of moisture content profiles. Note though, that within the zone of seasonal change of moisture content, considerable variations can occur at different times of year. Boreholes drilled in the autumn can determine the total extent of soil drying; boreholes drilled in the spring will only determine any residual persistent deficit
- Comparison of moisture content with index properties. For some over-consolidated clays with high liquid limit, there is an approximate relationship that the onset of desiccation occurs with a moisture content of less than 0.5 x liquid limit, with the onset of significant desiccation at 0.4 x liquid limit. Local correlations between moisture content and plastic limit may also be relevant in some situations
- Comparison of soil strength profiles. Desiccation of a clay soil causes increase in the undrained shear strength. As the fissuring of a clay soil affects its overall strength, these tests should be on small samples e.g., results obtained with a handheld penetrometer
- Effective stress or suction profiles

As each technique has its own advantages/disadvantages, one should use as many different criteria as possible, and only conclude that desiccation is present if the various tests corroborate each other. Of these tests, measurement of soil suction is usually the most reliable. However, it is also the most expensive, and, unless a comparison profile is determined away from all potential causes of desiccation, assumptions must be made about the equilibrium profile. Accuracy diminishes if samples are less than 90% saturated, and anomalous results can be obtained in clay which overlies a drained sand.

Refer to *Desiccation in clay soils*[54] for further information.

7.4.7 Assessment of shrinkability
The extent to which a clay soil will shrink or swell with changes in moisture content depends on many factors — most particularly the proportion of the various clay minerals within the soil (each of which differ in their properties), the proportion of non-shrinkable particles (those with a diameter greater than 425um) and the way in which the soil cracks to create a natural soil structure.

The plasticity index of the soil is generally used to provide a rough indication of the shrinkage behaviour, with high index indicating high shrinkage potential. This potential will be realised if the soil is dried over the full range from its liquid to its plastic limit, so there is a close correlation between shrinkage and plasticity index over this range. However, over the range of change in moisture content induced by vegetation, this correlation is not reliable.

Refer to *Low-rise buildings on shrinkable clay soils*[55] for further information.

7.4.8 Calculation of heave/shrinkage potential
Section 7.4.6 notes the issues with making a qualitative assessment of desiccation. Similar, and additional, problems occur for obtaining a quantitative measure of shrinkage. BRE Digest 412[54] describes the method of calculation, either based on moisture content or soil suction profiles. However, the results obtained are only estimates, and at best only

provide an approximate magnitude of any potential heave. In particular, both methods rely on using a water shrinkage factor to convert the moisture deficiency to vertical heave. This value is not constant with depth and will also differ with the shrinkage potential of the soil.

7.4.9 Assessment of permeability and rate of moisture recovery

The rate of a soil's recovery from desiccation is dependent on its permeability. The time period for a desiccated soil to return to full field capacity is dependent on the rate of recovery and the depth of desiccation.

Although tests are available for soil permeability, they are not usually appropriate for site investigations. Instead, a qualitative assessment must be made on the basis of examination of the soil conditions and soil structure. Large features such as bands of sand or gravel can provide means of rapid moisture movement and allow rapid rehydration. Smaller features, particularly the cracks which form between the soil peds, can also allow rapid moisture movement, particularly while the cracks are still open. Even after closure, the cracks are often lined with coarser sand particles, which improve soil permeability. A clay which has been disturbed so as to destroy its natural structure, e.g., puddled clay, may have an exceedingly low permeability.

7.4.10 Identification of root samples

Root samples taken from trial pits can be identified by specialist laboratories. This identification can indicate which species of tree have roots at a particular location. However, note that:

- The roots sent for identification are conducting roots, which are not themselves active in water uptake. Water uptake may be occurring at some other, more distant location
- Any water uptake may only be from the readily available water in structural voids in the soil, or may be adequately compensated by incoming water from other sources. Water uptake by active roots does not necessarily indicate that soil drying and shrinkage is occurring
- Some roots are far more obvious and easily seen than others. Sampling, particularly if only a few roots are taken, may not include all roots which are present
- Even if all roots are sampled, their proportion will not necessarily reflect their influence on soil moisture content
- Many roots are ephemeral and die. Ideally an iodine test should be used to check whether the root is alive, but even this test is not entirely reliable, as the root can remain alive in the soil for many years after a tree has been felled (with the root inactive throughout this period)
- Live roots may remain inactive in the soil, only becoming active in periods of drought

If root samples are taken, they should only be collected from below foundation level, and should ensure that all potentially relevant roots are sampled, i.e., from the full depth of observed root activity, and an adequate number taken to assess all species which might be present.

7.5 Possible contributing factors (and determining their involvement)

Various factors are particularly likely to cause or contribute to subsidence damage. The relevance of all such factors should be assessed. In the light of initial investigations, the need for more detailed investigations of particular aspects can be identified, and undertaken.

Diagnosis of the cause of damage should take account of (and be consistent with) all of the available information, or there should be a reasonable explanation to account for any anomalies. If there are any unexplained inconsistencies, additional investigations should be undertaken.

7.5.1 Trees

Although trees are a common cause of subsidence damage, their behaviour and interaction with soil conditions is varied and unpredictable. For this reason, no firm rules can be given on whether a tree, growing under any particular conditions or at a specified distance from a building, is, or might in future be, involved as a cause of damage (Table 8.1). Any tree beyond the distances given in Table 8.1 can probably be ignored, but the possibility of the involvement of trees at closer distances should be considered. However, just because a tree is at closer distance it does not follow that it will be involved as a cause of the damage.

Diagnosis of the involvement of a tree or other vegetation should be based on the crack or level monitoring (Sections 7.3.3 and 7.3.4) showing a clear pattern of seasonal movement. The monitoring should also have determined the extent and distribution of this movement. The identification of trees or other vegetation causing this movement requires detailed consideration. This information is best obtained by an arboriculturist, and should include:

- The proximity of the tree to the building, but noting that no definitive guidance on the risk of damage at any specified distance, is available
- The distribution of movement within the building. Localised movements imply a small tree or shrub in close proximity to the area of movement; more widespread movements imply a larger tree capable of exerting an influence over a large area. Movements are normally greatest close to the tree, but variations in soil conditions can produce abnormal results
- The identity of the tree. Some genera of trees are considered to be more likely than others to cause soil drying and damage (Chapter 8). However, as there can be considerable variation between individuals, any classification of this sort can only provide general guidance
- The crown volume of the tree (crown height and spread), as a general indication of the total leaf area. The greater the leaf area, the greater the potential to lose water from the leaves during transpiration. Crown volume is far more relevant than overall height
- The vigour of the tree, determined either from its shoot extension growth or from the amount of tissue laid down in the annual growth rings. This provides an indication of the tree's overall efficiency and whether it is transpiring at its maximum potential
- The history of the tree. This includes consideration of its age, previous growth rate, previous tree surgery work such as crown reduction or pollarding, or damage to the root system. The overall distortion in the property (determined from the distortion survey, Section 7.3.2), should be consistent with any potential previous involvement of the tree
- The depth of foundations determined from trial pits (Section 7.4.1) and vulnerability of the building to damage
- Root identification (Section 7.4.10)

7.5.2 Slope

Slopes can be unstable, particularly if the underlying soil is clay. Any form of slope instability will cause a progressive downward movement, possibly combined with lateral movement of the foundations. These movements should be detectable by level and crack monitoring, and by a distortion survey.

If the site is sloping and instability is suspected, the following information should be sought:

- Full details of the subsoil to a depth of at least 6m
- Measurement of the angle of the surface of the clay (which may not correspond to the soil surface)
- Any change in conditions or increased load on the upslope side
- Any change in condition or excavation on the downslope side

7.5.3 Drains

If leaking drains are suspected, the distortion survey should reveal subsidence in proximity to the drain, and level and crack monitoring should show progressive movement. If movements of this sort are suspected, the condition of the drain should be checked. CCTV scans provide a reasonably cheap and effective method. A water test will determine the rate of leakage along the pipe.

Many old drains may have deteriorated and leak, without necessarily causing subsidence damage. The condition of these can also be checked by CCTV and water tests.

7.5.4 Traffic, or other causes of vibration

It is unusual for vibration to cause significant damage. If it does occur, there should be a close correlation between the development of damage and the occurrence of the vibration.

7.5.5 Other building operations

Any changes within a building may cause damage. This could include:

- Changes in temperature or humidity, causing thermal or volumetric alterations in materials
- Structural alterations, causing changes in foundation load, or deflections in the superstructure
- Additions, causing initial settlement of foundations

Any damage of this sort can usually be traced to the changes that have occurred.

7.5.6 Changing environmental factors

The properties of some types of ground e.g., sands, clays and peat can be affected by changing water levels, whether a result of natural occurrence, services construction or abstraction activity.

Consideration in any investigation should be given to the risk of damage that could be caused by these altered states e.g., the 'French drain effect' (Section 5.2.6.1) through backfill to service trenches, changes in water abstraction activities etc.

7.6 Final report

On completion of the investigations, a report should be issued. This should:

- Identify the cause of damage. The cause should be consistent with all of the results, and sufficient information should be available to ensure that the correct cause has been identified
- Consider the future prognosis if no remedial action is taken
- (if remedial action is required) Identify the options and their technical merit
- Estimate the costs of these options
- Identify whether any other action is required, e.g., the seeking of further specialist advice

The Expert will normally project manage the necessary remedial action (Section 9.4). On completion, the Expert should be able to give the Owner an assurance that in their professional opinion, the property will remain structurally stable and free from significant movement in the future. The assurance would assume no changes in the property's environment or use and also that the property will be reasonably maintained. The assurance would be a considered professional opinion and not a warranty or guarantee. It can take the form of a 'Certificate of Structural Adequacy' (Section 10.8 and Appendix B).

Such an assurance, when given, should encourage Lenders, Insurers and Valuers to accept the property with little if any effect on value because they will have the benefit of knowing that an in-depth assessment of the problem has been made and that the structural relevance of the matter has been objectively evaluated.

8 Trees and tree management

8.1 Introduction

8.1.1 Nature of damage and tree physiology
Direct damage to structures by physical growth of the trunk, branches or roots of a tree, is often very easy to diagnose by observation (Section 5.3.14).

Shrinkage and swelling of clay soils causes damage to structures (Section 5.2.3). This situation can be exacerbated by the presence of a single tree, multiple trees, hedges or other vegetation. This chapter provides advice on steps to be taken to identify contributory action from roots, and the methods to deal with such issues. Drain blockage by roots is considered in Section 8.2.4.

Climate change has become an increasingly important factor in the assessment of the tree/foundation relationship, and the decision to remove a tree (or manage a tree, if deemed appropriate) needs to consider the amenity and environmental benefits of the tree e.g., shading, transpirational cooling, biodiversity, and health and wellbeing provided to the population. The Trees and Design Action Group's (TDAG) *First steps in valuing trees and green infrastructure*[56] outlines the strengths and weaknesses of different valuation systems for both amenity and ecosystem services values.

A tree's water-use and its rooting extent is not a function of its height. In 2004, a report by East Malling Research and Cambridge University, known as 'Hortlink 212'[57], established that *crown reduction* did control a tree's water uptake, whereas *crown thinning* was not as effective. Ten years later, a pilot study known as 'Beyond Hortlink 212'[58] determined that crown reduction, when carried out regularly, was effective at reducing incidences of damage in areas of risk.

Every UK subsidence event where it is suspected that a tree or other vegetation could be implicated, must be treated on a case-by-case basis using the evidence-based research published in the Kew Tree Root Survey that was commissioned immediately after the Great Storm of 1987. An analysis of the survey results was published by Cutler et al in 1990[59], and Table 8.1 in this guide presents some of the key information.

8.1.2 The tree: crown-root balance
Trees are normally long-lived. During their lifespan they pass through three phases:

- The juvenile phase, of relatively rapid growth
- A mature phase, where the rate of growth slows
- A phase of senescence, involving slow decline, with very reduced growth in some parts, and dieback in others

The size of the trunk and crown are in balance with the root system. Damage to either affects the other. Removing a small part of the crown may lead to moderate regrowth. Cutting out a large proportion may, among other things, permanently stunt the tree. It could cause physical instability if cutting is unbalanced, or stimulate vigorous and often juvenile regrowth. The response to crown cutting depends on the species and the season during which the crown is cut. Cutting roots without compensating by reducing crown area can put the tree under water stress. Great care should be taken if it becomes necessary to cut roots. Loss of stability, or entry of disease organisms may also ensue.

8.1.3 Water use and pruning
The trunk, branches and twigs provide a means of exposing leaves to sunlight, away from the shade of smaller plants.

Leaves use water and minerals provided through a plumbing system that extends from the roots through the trunk, along the branches and twigs. Water in the leaves combines with CO_2 from the air, using energy from sunlight and the agency of chlorophyll, producing substances that are used to build cells and tissues and their contents.

This plumbing system is renewed or added to each year by a division of cells (vascular cambium) to the inner side of the bark. In temperate regions this gives rise to growth rings. The new wood and phloem (inner bark) formed add to the tree diameter. Phloem cells transport soluble synthesised materials. If large areas of bark are damaged, down to the wood, the water flow can be seriously disrupted (and there could also be problems in moving soluble food reserves between root and crown).

Water vapour passes through regulated holes in the leaves (stomata) to the air and in so doing provides part of the power enabling flow of water to occur from the roots through the tree. This mechanism is in part driven by a negative water potential gradient from the soil/root surface interface to the leaves. The heat loss through this transpiration helps keep the leaf within its functional temperature range in hot weather. The size of the stomatal pores is under the control of the tree, and plays a part in regulating the rate of water movement.

Given adequate water supply and good drying conditions during the day, the amount used by a tree is closely related to the total leaf surface area. Different tree species have different water requirements.

These are often reflected in the differences that can be seen in leaf size, shape and texture. Some species of tree have the ability to close stomata and stop transpiration at maximum peak midday temperatures regardless of the availability of water at the soil/root surface interface. There is no direct connection between the height of a tree and its water uptake. Overall leaf surface area (i.e., crown volume and crown architecture) dictate water uptake. Research is ongoing into which species control their water uptake best in periods of drought.

It follows that removing branches temporarily reduces the evaporating surface area of leaves supported by a tree, and hence reduces water usage. Regulation of water use by trees can be achieved through pruning.

Research has shown that a particular type of pruning can assist in reducing transpiration and therefore water uptake. It does so by preserving the leaf boundary layer in a denser smaller crown which reduces the through-flow of air which then lowers transpiration[57, 58, 60–62].

Inappropriate removal of twigs and branches may lead to undesirable rapid compensatory regrowth. Expert arboricultural advice is always needed, since effective methods of management vary from species to species.

8.2 Roots

8.2.1 General

Roots give both mechanical support for the tree and take water and dissolved minerals from the soil.

Much of the mechanical support is provided by the larger roots. In many species, these 'large' roots are themselves surprisingly narrow (with even the largest often not exceeding 10cm in diameter at 1m from the trunk). They are also much shallower than might be expected. Roots may be classified as either 'structural' (the thicker ones) providing mechanical support, or narrower 'conducting' roots. To a degree, conducting roots still contribute to a tree's stability.

The Kew Tree Root Survey reported that from a sample of 760 individual trees whose root diameters were measured at 1m from the trunk:

- 4% had only fibrous roots
- In 23% of the sample, the thickest roots ranged between 2–5cm in diameter
- A further 40% of the sample had thickest roots in the range of 5–10cm
- Roots over 10cm in diameter occurred in only 33% of the sample

This means that in general, structural roots forming the main framework of the root system are relatively narrow. They might be dismissed as of little significance and cut while working in the ground. This could lead to instability in the tree.

Further observations were made on root depth of a sample of 3,469 trees:

- In 15%, the structural roots were less than 0.5m deep
- In 31% the roots were between 0.5–1m deep

- Just over 50% were found at a depth of 1–2m
- Only approx. 4% had roots extending deeper than 2m

The common perception is that the larger roots are normally deeper than the data indicates. Again, there is the possibility that important structural roots may be severed without realising the seriousness of the consequences. Conducting roots are responsible for much of the water uptake. They also provide some mechanical support, frequently extend to greater depths than the larger roots and occupy very large volumes of soil. They require an oxygen supply to respire.

Compacted and waterlogged soils provide poor rooting conditions and limit areas where roots may occur in the soil.

Roots of a solitary tree in an open level field can be expected to radiate fairly evenly from the trunk (more so in conifers than other trees) in good, uncompacted soil.

Generally, it is almost impossible to predict where the roots may be growing. This is because their distribution is influenced by competition from other nearby trees and shrubs, variability in soil conditions and slope. They are also affected by the direction of water supply and prevailing wind direction. In urban settings root growth is limited by physical obstructions and compaction (Figure 8.1).

Figure 8.1: Typical tree root systems[63]

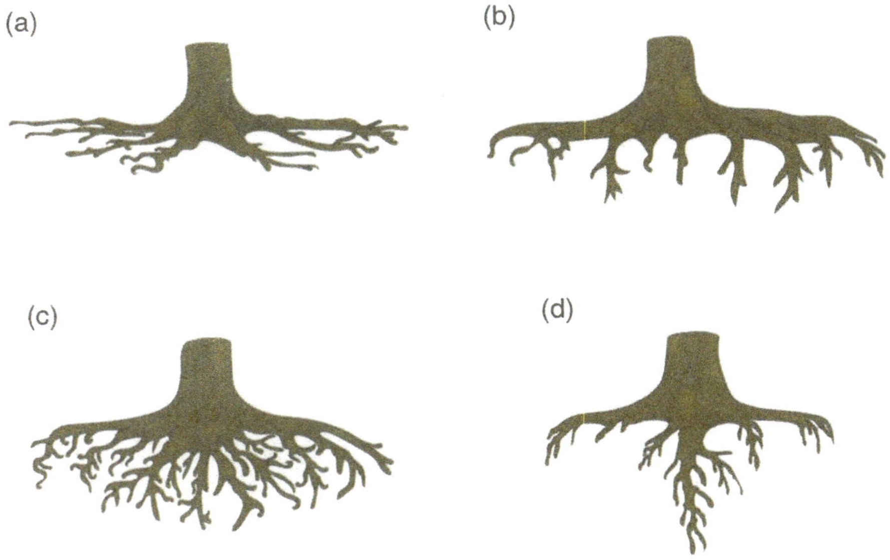

(a) Plate root systems – are dominated by lateral roots growing at a relatively consistent depth below the soil surface. Eg. fir (*Abies spp.*), spruce (*Picea spp.*) and beech (*Fagus spp.*).

(b) Sinker root systems – have a plate of lateral roots that develop vertical sinker roots.
Eg. Some oaks (*Quercus spp.*)

(C) Heart root systems – display a central complex of sinker roots as well as oblique and shallow lateral roots giving a more compact rooting structure.
Eg. Birch (*Betula spp*), larch (*Latrix spp.*), oaks (*Quercus spp.*) and lime (*Tilia spp.*).

(d) Tap root systems – maintain a persistent positively geotrophic root from the radicle in addition to developing a plate of shallow lateral roots.
Eg. Many oaks (*Quercus spp.*), pines (*Pinus spp.*) from Mediterranean climates.

NB. Tree root morphology can vary depending on the soil and tree species interactions. There can be genetically driven tendences, but these certainly do not result in fixed morphological types. For example, European oaks tend to have a heart-root system, but this is not absolutely true in all scenarios. So be aware that root systems of tree species cannot always be categorised.

Source: Hirons, A.D and Thomas, P.A. (2018). Reproduced with permission of John Wiley & Sons Ltd

There is no simple relationship between tree height and root spread. Small trees with dense crowns may have a wider root spread than taller trees with sparse crowns. In all the survey results analysed at Kew in preparation for *Tree roots and buildings*[64], the authors found no correlation between root spread and tree height for any of the species studied.

8.2.2 Interpretation of root spread data

The data from the Kew Tree Root Survey should be interpreted with care. Practically all of the data collected were for trees growing in shrinkable clay soils, where it was thought that the tree might be responsible at least in part for some subsidence.

These trees which have caused damage are a very small proportion of the overall tree population, the vast majority of which are growing as close or closer to buildings with no apparent consequential subsidence.

The Kew Tree Root Survey data for maximum recorded root spread indicates that at any greater distance it is very unlikely that a tree would be involved in damage to buildings (Table 8.1). The original data show that, in the vast majority of cases, the trees implicated were very much closer to buildings than the maximum recorded distance. Very few trees from each sample were at or near the maximum distance.

The distance figures within which, for example 50% or 75% of trees of a particular species were recorded as being implicated in damage are just that — they do not represent percentages of 'risk' of damage.

It might be assumed, that by taking the figure that accounted for 90% of the recorded instances as a reasonable distance for that sort of tree to be from a low-rise structure, one was accepting a 10% increase in 'risk' of damage. A moment's consideration of the aforementioned points will show the fallacy in this argument. Added risk is very difficult to assess. In this example it would be a small fraction of 10% at most. The data provide an historic record of what has been found. Although they give some guidance as to whether a particular type of tree might be implicated in subsidence, to be certain a proper investigation is always needed. Equally, it is unwise to use the data predictively.

They may be used prudently when selecting trees for particular circumstances, but do not prescribe how or where the roots will grow, particularly in the context of built-up areas.

An essential element in the process of assessing whether current damage might be in part due to a particular tree or trees is to find and identify roots (using DNA testing if necessary) from in an inspection pit at the region of failure of the structure (Section 7.4.1). If it has been proved that a tree is partly or totally the cause of subsidence damage, other features of the tree and the position in which it is growing should be taken into account in assessing the necessary remedial treatment.

In the case of valuable trees or where there is more than one species on a site, consideration should be given to the DNA testing of roots — which can potentially identify which tree the identified roots originate from.

8.2.3 Embodied carbon, net zero requirements and the role of trees

Produced more than 35 years ago, the Kew survey did not include the increasing impacts of climate change, the requirements of the *Climate Change Act 2008*[29] and the role of trees in storing embodied carbon as a contributor to achieving net zero.

The issue of carbon and the commitment to achieving net zero will be an influencing factor in determining whether to retain trees if they are proved to be a factor in causing structural damage. A carbon calculation for the subsidence repair should be carried out along with the cost-benefit calculation to determine the optimum solution.

Note that every tree contributes to the urban forest and the resilience of the urban forest is a key component of responding to climate change.

8.2.4 Roots and drains

It is improbable that roots penetrate drains that are watertight and have no leaking joints. If water condenses on cool drain pipes, roots may be attracted and give the appearance of 'attacking' a drain. Uneven changes in soil volume caused by roots may bring about movement of drain pipes, causing them to leak. Replacing leaking systems with modern drains, using flexible joints where necessary will normally solve the problem. Consequently, unless the tree from which the roots grew is likely to become involved in other damage, it needs no remedial treatment provided drains are repaired using appropriate materials.

8.3 Remedial action

The site investigations described in Chapter 7 should have identified:

- Whether trees are a cause of damage
- If they are, which tree or trees are involved

- Where trees are involved, one option for remedial action is to deal with the cause so as to restabilise the foundations. This is usually a cheaper option than underpinning, but, if it is to be effective, it must be properly implemented with a full understanding of the objectives and methods. Arboricultural advice is likely to be essential

Remedial action with trees can only be effective if the influence of the tree on the soil is primarily seasonal, i.e., if the soil rehydrates fully, or to near field capacity, each spring. If there is a deep-seated persistent moisture deficit, any action to reduce the influence of the tree will allow the soil to rehydrate and swell, causing heave of the foundations. It is not feasible to try to 'balance' the moisture content so as to prevent both heave and subsidence. Recovery from a slight persistent deficit, which is complete within 1–2 years, will probably be acceptable. However, note that heave recovery can be beneficial in many instances by closing recent cracks, but if recovery is going to take too long and continue to worsten the damage, underpinning with full anti-heave precautions is the only viable option. No form of tree management will be effective in such instances.

For this reason, site investigations must identify:

- Which trees are involved, so that these can be targeted for remedial action
- Whether there is a persistent moisture deficit to indicate whether heave is likely

Table 8.1 presents some of the information gathered by the Kew Tree Root Survey. The records relate to trees which were thought to be connected with damage to buildings. They are historical and should not be used predictively. The distances given in Column 3 are based on various factors:

- They take into account the percentage of cases of all damage recorded at such distances for trees of that species, and the relative frequency of planting the species of tree in relation to the frequency of damage reported in the survey for such trees
- The natural vigour of the type of tree in question has also been taken into account. Just as maximum root spread figures can be readily misapplied, so can these — they do not constitute a management tool
- The figures should only be used to assist in planning the next steps of an investigation leading to determination of the appropriate action/treatment
- The figures should never be used to decide whether a particular tree should be felled

Table 8.1: Records for trees that have been thought to be implicated in damage

Tree species	Normal mature height in clay soils in urban setting (m)	Adjusted distances (m) (Section 8.3)	Notes relating to trees recorded in Kew survey
Apple, *Malus*; Pear, *Pyrus*	8–10	4–6	Smaller fruit trees can be managed readily, even closer than indicated
Ash, *Fraximus*	23	10	Roots often very shallow; crown tolerant to pruning
Beech, *Fagus*	20	9	Rarely implicated in damage; scarce on clay; old trees may become unstable because of root rot
Birch, *Betula*	12–14	4–5	Low implication in damage; older trees tolerate only light pruning
Cypress, *Cupressus, Chamaecyparis*	15–25	3–4	Soil close to trees dries excessively; cast shade and may be undesirable at these distances
Cherry, *Prunus*	12	3–6	6m for vigorous varieties; very shallow roots; tolerates only light pruning
Damson, *Prunus*	6–8	3–6	Shallow rooted; tolerates only light pruning
Elm, *Ulmus*	20–25	12–19	12m for those persisting in hedges limited in height by Dutch elm disease
False Acacia, *Robinia*	18–20	10	Initially fast growing; older trees tolerate only light pruning; drops branches

(continued)

Table 8.1: *continued*

Tree species	Normal mature height in clay soils in urban setting (m)	Adjusted distances (m) (Section 8.3)	Notes relating to trees recorded in Kew survey
Hawthorn, *Crataegus*	10	5–7	Soil close to hedges may dry excessively; tolerates heavy pruning
Holly, *Ilex*	12–14	3	Rarely implicated in damage; tolerates heavy pruning
Horse Chestnut, *Aesculus*	16–25	10	Tolerates heavy pruning and readily managed at closer distances; may shed branches
Laburnum, *Laburnum*	7–9	7	May be short lived; rarely implicated in damage
Laurel, *Prunus laurocerasus*	8	3–6	Tolerates heavy pruning
Lime, *Tilia*	18–24	8–11	Manageable at closer distances, tolerates heavy pruning
Magnolia, *Magnolia*	5–7	5	Rarely implicated in damage
Maple, *Acer* species (Sycamore)	17–21	12	Refers to larger species; tolerates heavy pruning and can be managed at closer distances; smaller species or varieties may be grown much closer to buildings
Oak, *Quercus*	16–23	18	One of the more problematic genera
Pine, *Pinus*	20–29	8	Rarely implicated in damage
Plane, *Platanus*	25–30	8–10	Tolerates heavy pruning and can be managed at closer distances
Plum, *Prunus*	6–8	3–6	May be shallow rooted
Poplar, *Populus*	25–28	20	One of the more problematic genera; roots close to trees can be shallow; tolerates heavy pruning
Spruce, *Picea*	15–18	7	Rarely implicated in damage
Sycamore, *Acer pseudoplatanus*	20–24	12	Often seeds into inappropriate places; tolerates heavy pruning
Walnut, *Juglans regia*	12–15	8	Rarely implicated in damage
Whitebeam/Rowan, *Sorbus*	8–12	5–7	Relatively unproblematic; tolerates heavy pruning
Willow, *Salix*	15–24	18	One of the more problematic genera; tolerates heavy pruning
Yew, *Taxus*	8–12	5	Rarely implicated in damage

Note: Derived/adapted from *Tree roots and buildings*[64].

Tree Preservation Orders and Conservation Areas

Before any work is undertaken it is essential to determine whether a tree is subject to a Tree Preservation Order (TPO) or is in a Conservation Area. The Local Authority can advise. In some situations, if it can be established that the tree is a 'nuisance' in the legal sense, exemptions may apply. Otherwise, nothing should be done without the written agreement of the Local Authority, and then only agreed work should be carried out. If a tree is the cause of damage, and consent to take necessary action is refused, the Local Authority may be liable to pay compensation for loss or damage caused or incurred in consequence of that refusal. This is set out in Section 203 of the Town and Country Planning Act 1990[65] and Regulation 24 of the Town and County Planning (Tree Preservation) (England) Regulations 2012[66].

There are two main options for remedial action:

1. Fell the offending tree/s
   This may not be necessary, and indeed may be detrimental to the environment and the value of the property. In the case of trees of high amenity value[67] it should be considered as the last resort. Felling will immediately stop all further root activity. The roots may take several years to die, and could give rise to sucker shoots, but they will immediately cease all water uptake unless suckers are allowed to develop. The soil will start to rehydrate. Provided the influence is primarily seasonal, rehydration will usually be complete by the spring after the tree was last growing, i.e., if a tree is felled in the spring, moisture recovery will already be complete; if felled in the autumn, rehydration will be complete by the following spring. If there was a persistent soil moisture deficit, recovery will take longer, e.g., the winter of 1975/6 was unusually dry and aggravated the effects of the dry summers either side of it. The impact of felling a tree that predates the building should also be considered in light of unintended consequences such as heave[68–69].
2. Prune the offending tree/s
   The objective of this treatment is to reduce the leaf area by reducing crown volume and reshaping the crown architecture of the tree, and therefore its ability to transpire and lose water. If the extent of seasonal soil drying is reduced, it will reduce the amplitude of seasonal movement[60]. It is necessary to reduce this amplitude to below that which can be tolerated by the building, so that crack movement is stopped or kept to an acceptable level. There are potential difficulties:
   - The amount of pruning must be adequate to prevent unacceptable movement, while avoiding excessive work to the tree
   - Most trees will respond to heavy (or even moderate) pruning by vigorous regrowth which restores the previous leaf area and crown volume. It will therefore be necessary to repeat the treatment before an unacceptable amount of regrowth has occurred, and to continue this treatment as long as the tree remains. In principle it is recommended that tree management in urban areas on clay soils should occur every 2–3 years depending on arboricultural advice. This imposes a continuing responsibility on the Owner to organise and finance the work
   - The tree must be amenable to the extent of pruning which is required. The tree must continue to be a benefit i.e., providing ecosystem services as well as amenity. Inappropriate pruning irrevocably spoils many trees
   - In rare situations other options such as root barriers, root severance, or artificial control of soil moisture may be viable (Section 8.1.2). However, these methods are usually less reliable, harder to implement, and may have undesirable side effects. Specialist advice should be taken before considering such methods

Arboricultural expertise is required when deciding whether to prune or to fell.

This should consider:

- The ecosystem service value and amenity of the tree[56], or other derived values, such as health and wellbeing. With a tree of very high value, it may be better to accept the cost of underpinning, so as to retain the tree and its associated values, depending on the approach taken by Insurer and the scope of the policy. Conversely, a tree of low value is probably best felled, recurrent costs of pruning will not be justified
- The ability of the species to tolerate the necessary amount of pruning. Some species, for instance London plane are very amenable, whereas others are difficult to manage under a reasonable pruning regime
- The health, condition and vigour of the tree as an individual, and its response to treatment
- The practicality of ensuring future maintenance

If pruning is adopted, arboricultural advice is required to determine the most appropriate method and its extent. This may involve one of the following:

Crown reduction
Reducing the overall volume of the crown by a specified percentage and altering the crown architecture to reduce the through-flow of air (Figure 8.2).

Pollarding
Extreme shortening of main structural branches. Generally speaking, pollarding is a street tree management approach and used in regular pruning regimes in areas of risk (Figure 8.3).

Figure 8.2: Crown reduction

Figure 8.3: Pollarding

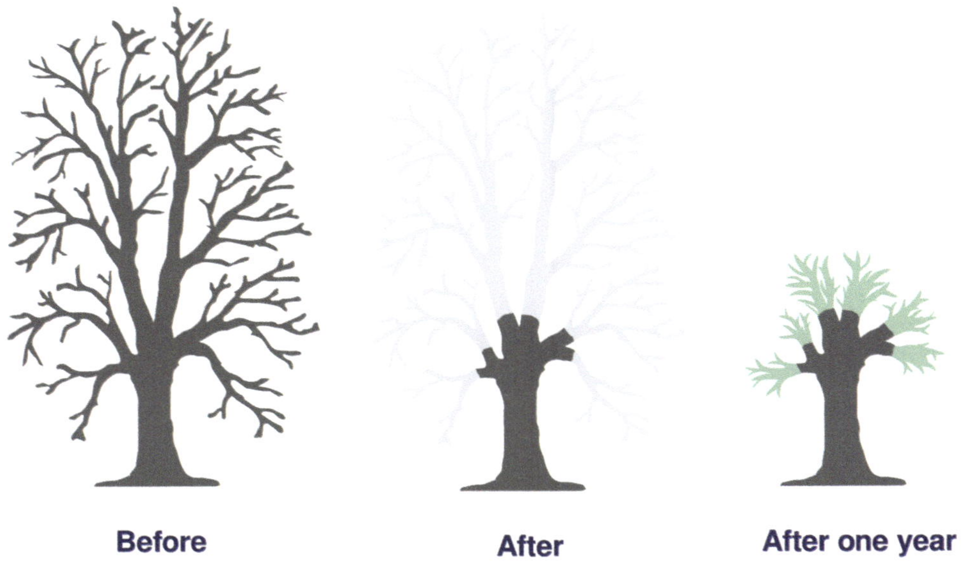

Crown thinning (no longer recommended)

Selectively reducing the density of branches by a specified percentage, while retaining the overall height and branch spread (Figure 8.4).

Note: Hortlink 212 has proven that crown thinning is ineffective at controlling a trees' water use because it increases through-flow of air in the crown. It should not be used as a remedial measure for subsidence.

The choice of crown reduction or pollarding will depend on the species of tree, its reaction to the different methods, the severity of pruning required, and the options for future management.

The choice should depend on individual circumstances, and so no definitive guidance can be given.

Figure 8.4: Crown thinning

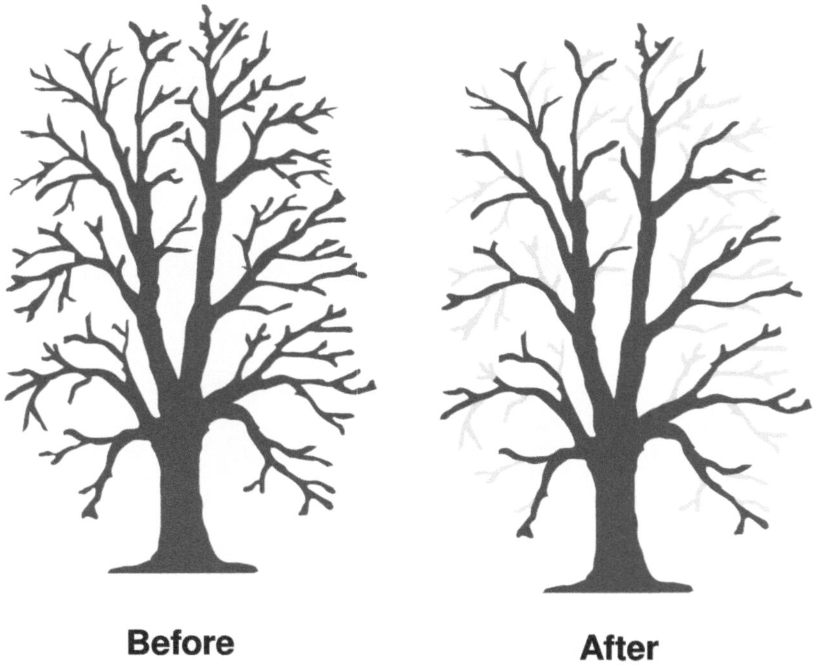

If something must be done but none of these methods is appropriate, it may be better to replace the tree with a more suitable species.

Likewise, no definitive guidance can be given on the severity of pruning:

- Each case must be taken on its own merits and determined by an arboriculturist
- In general terms the leaf area/crown volume must be reduced sufficiently to ensure that, by the time the tree has regrown and next requires pruning, its leaf area/crown volume is still significantly below that which initiated the damage
- A balance needs to be struck which takes account of the severity, the time interval before the tree is next pruned, the ability of the tree to tolerate the regime imposed, and the appearance and functionality of the tree after pruning. However, it must be sufficient to ensure that adequate stability of the foundations is achieved, and can be maintained in subsequent years

Those trees in the vicinity that are not involved in damage need not be pruned unless there are sound arboricultural reasons to do so as assessed by an arboriculturist.

Where they have adequate space and are not interfering or are unlikely to interfere with buildings, all that is needed is a regular, documented check for safety and disease, followed by appropriate action. The duty of care of the Owner is thus discharged.

Engineering solutions to walls to accommodate trees

Trees of high value enhance property values and should be accommodated where possible. Where such trees are causing damage to small structures such as walls, garages and unattached outbuildings, engineering solutions should be considered such as bridging or adjusting walls. There are various ways to bridge walls:

- With a fence infill (Figure 8.5a)
- As shown in Figure 8.5a, but with railings
- A lintel can be placed away from the tree roots and the wall built above it, should continuous walling be required (Figure 8.5b)

Figure 8.5a: Bridging a wall with fence infill

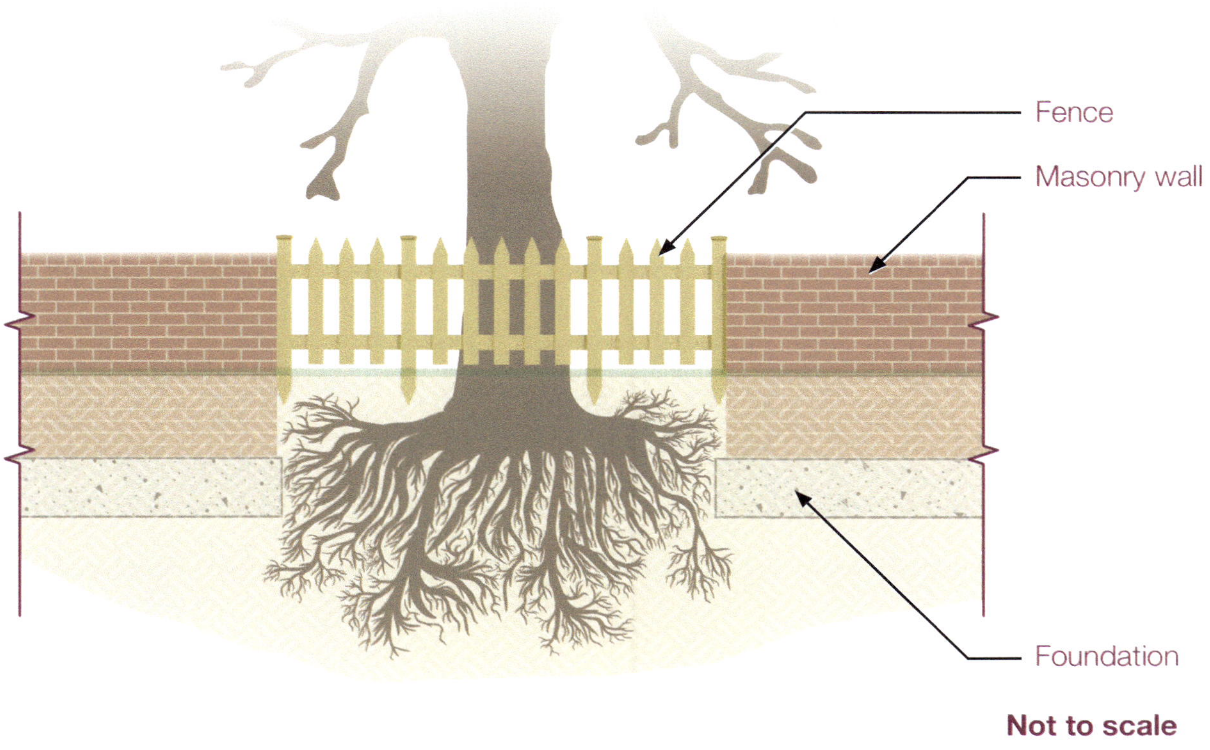

Figure 8.5b: Bridging a wall with a lintel

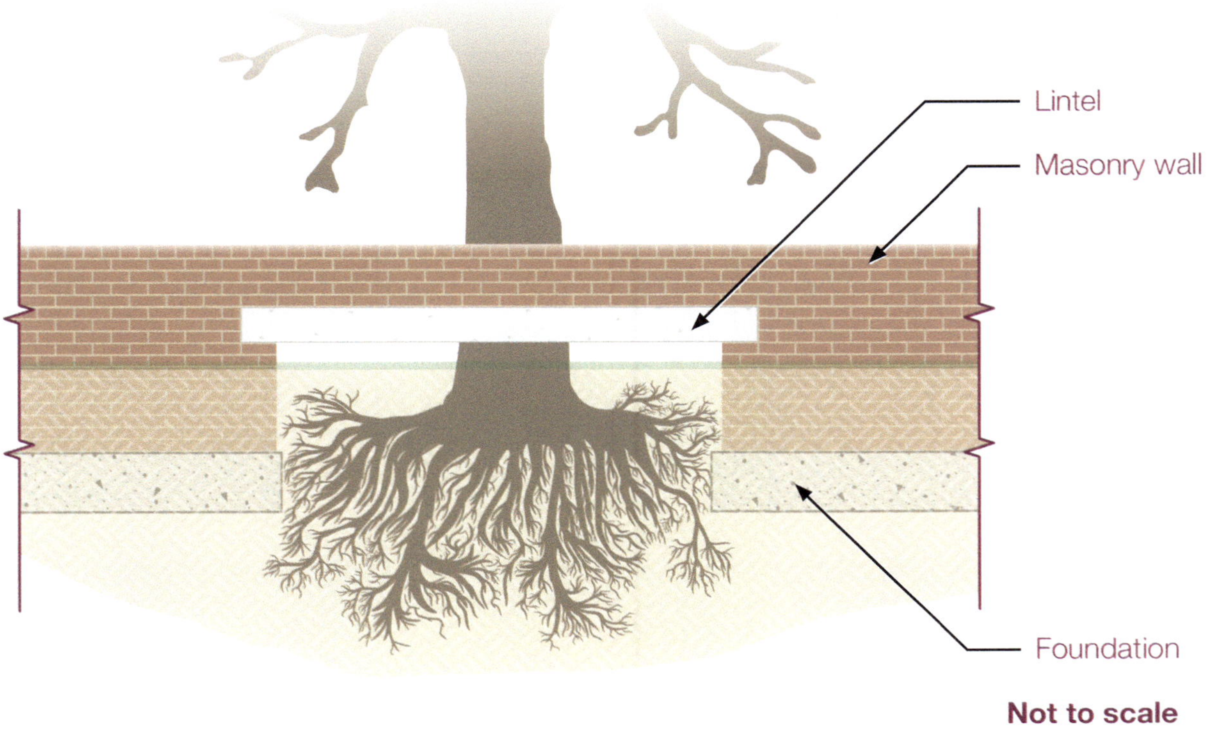

8.4 Root barriers

The use of root barriers should be avoided if possible. Root barriers are not a 100% effective solution. At best they are an intermediate remedial operation which may be breached or degrade over time. Another consideration may be the creation of instability if roots on one side of the tree are substantially removed to accommodate the barrier, as well as the likelihood of growth over and under the barrier. To be as effective as possible, they must extend for the full distance of possible root spread. Roots have been known to grow through backfill and under a barrier of 2.8m depth or, indeed, over the barrier. If used, they need to extend above ground level. Roots have been seen to grow above barriers and re-enter the ground on the far side. Points at which barriers are breached for services may allow root penetration.

Root barriers may alter groundwater flow, which can give rise to other problems. Microporous barriers are not yet fully tested.

When inserting root barriers, care must be taken to minimise damage to roots.

8.5 Monitoring after remedial action

Regardless of whether a tree has been felled or pruned, it is good practice to monitor it, to confirm that the action taken has been effective in restabilising the foundations. If the tree has been felled, there is the need to confirm that significant heave does not occur and that no other trees and vegetation are continuing to cause movement. If the tree is being pruned, monitoring is required to demonstrate that the amplitude of movement has been reduced to an acceptable level.

Monitoring should be continued through at least one summer after pruning. For both pruning and felling, level monitoring (Section 7.3.4) should be the preferred option. It is particularly useful if level monitoring has commenced prior to the action on trees, so that one can compare movements occurring before and after the action of trees (monitoring recovery movement through the winter can establish the extent of seasonal movement occurring; there is no need to monitor through a growing season before taking action with trees unless it is to assist in implicating the tree as the cause of the damage in the first instance). In addition, it may be beneficial to monitor cracks to confirm that the stabilisation of the foundations has also been effective in preventing further significant crack movement.

Once adequate stability has been confirmed, superstructure repairs and redecoration can begin (Section 9.6).

If trees are being pruned, it is desirable that level monitoring markers should be left in place as a means of checking on the adequacy of future tree management.

Once it is confirmed that the foundations have been adequately stabilised, it may be appropriate to issue a report which:

- Identifies the original problem
- Identifies the remedial action taken
- Specifies the future management regime (if trees have been pruned)
- Confirms the action taken has been effective in stabilising the foundations

The history of level monitoring will be an invaluable part of such a report.

8.6 Tree management for prevention of subsidence damage

There are no sound economic grounds for widespread and indiscriminate pruning of trees in case their growth could lead to subsidence in the future. The cost of such work would far outweigh the cost of remedial treatment in those instances when direct connection between subsidence damage and trees has been established. However, targeted cyclical pruning in areas of known incidences of damage is a prudent management option. Individual circumstances are paramount, and the possible interaction of each tree and building should be considered on its own merits, as part of the general care of the tree.

8.7 Tree law

There are statutory protections or considerations that apply to the felling or management of trees. This may include the requirement to obtain a felling licence from the Forestry Commission under the Forestry Act 1967[70] if more than 5m^3 of timber is to be felled in one calendar quarter (or 2m^3 in a calendar quarter if the wood is to be sold — for logs, for example). This applies to trees in hedges as well as woodlands. Trees covered by TPOs or within Conservation Areas are also protected, and authorisation must be obtained from the local planning authority pursuant to the Town and Country Planning Act 1990 before any work can be carried out. For the purpose of the protection, 'work to the tree' includes work to the roots as well as to the above-ground part of the tree. Any damage done to a protected tree's roots without permission of the local planning authority may be a criminal offence. Any impacts on the tree's root system requires the Local Authority to approve any works undertaken.

Other interested parties may include the local Highways Authority (usually the council), third-party Owners of protected trees and other interested groups such as residents' associations and tree protection groups. Also, for the sake of completeness, the Owners of underground utility infrastructure can also be affected. Water mains, drainage, electricity, gas, telecommunications conduits and tree roots can be at risk of damage from pressure injection of substances into the ground in their vicinity. As statutory undertakers, the likes of the gas, electricity, and telecommunications cables/pipes are laid pursuant to statutory powers, and the injection might interrupt the execution of such powers.

Where there is an allegation that a property has suffered from tree root subsidence, there are possible legal claims which might follow the event.

In the pre-litigation stage of a claim, the Joint Mitigation Protocol[71] advocates conduct between the Homeowner and the tree Owner, where it is a Local Authority-owned tree causing alleged damage. This protocol is undergoing review and reference should be made to the current version when considering its use. It sets out an agreed timeline for communication and a set of investigation results which can be used to speed up a claim in the 'mitigation' phase where works to the tree are being considered to limit the scope of required repair works.

Primarily there is a potential civil claim in nuisance/negligence against the tree Owner/tree controller. This is a claim based on the alleged breach of the duty of care owed by the tree Owner/controller to ensure the tree did not cause damage to neighbouring properties. The law which determines this claim is set out in the steps of proving a negligence claim and in previous decisions of the Civil Court. One point to note is that it is not simply enough to evidence that the tree caused damage to the property. Issues such as whether the tree Owner/controller was able to foresee the risk of the tree causing damage to the property before the damage occurred, as well as issues of contribution, should be considered. Legal monetary damages which can be pursued to compensate the Owner of the damaged property must be reasonable. There are likely to be a range of reasonable remedial solutions to the property and the damages claim must reflect those reasonable options, even if more money was actually spent on a repair for other reasons. The law in this area is complex and can change as new case law emerges. Legal advice should be sought should a claim in nuisance/negligence be considered.

There is also a potential claim pursuant to Section 202E of the Town and Country Planning Act 1990 — if the cause of the subsidence damage was a tree protected with a TPO. If evidence shows that removal of the protected tree would stop ongoing movement to a property and the planning authority refuse an application to remove/reduce the tree, there is a possible claim for compensation against the authority for the additional costs incurred as a result of that refusal. This is a legal claim pursued in the Upper Tribunal Lands Chamber. The Town and Country Planning Act 1990 contains the legal framework for this claim but previous decisions of the Lands Chamber also impact on how such a claim will conclude.

Another important point to consider is a possible claim for a legal injunction. This is largely pursued as a first step in a claim for damages in nuisance/negligence. Where a tree is alleged to be causing continuing subsidence damage and the tree Owner/controller refuses to reduce or remove that tree, and tree removal is found to be the most appropriate remedial solution, a claim to seek an award from the Court to force the tree Owner to carry out the works to the tree can be considered. A note of caution should be raised here as these 'applications' can take a significant amount of time and can have limited success. Previous Court decisions should be referred to when deciding whether to take this step. To date Courts have been reluctant to order a draconian injunction where insurance cover is in place which would cover an alternative repair solution to tree removal.

In respect of repair works, the Access to Neighbouring Land Act 1992[72], should be considered. This is an Act which can help where a third party refuses access to their land which might be required for repair works to go ahead to the damaged property. If there is no alternative method of carrying out repair works, and the third party refuses access to their land, an application can be made to the Civil Court for an order compelling the refusing Landowner to grant access. This can have cost implications in that the inconvenienced Landowner should be compensated but it can provide a solution in some claims where repairs could not otherwise go ahead.

Aside from claims arising from the damage itself, there could be other possible legal claims in:

- Breach of contract — possible claim against Insurers if a claim is denied and that denial is disputed
- Professional negligence — e.g., if subsidence is missed on a pre-purchase survey and a claim is to be considered against a Surveyor
- Fraudulent misrepresentation — e.g., against a previous Owner of a property
- *Defective Premises Act 1972*[73] claim — e.g., a claim against anyone involved in the construction of a modern dwelling which led to a failure of quality
- Breach of the *Party Wall Act 1996*[14] — for breach of a party wall award

This list is not exhaustive and there could be other possible legal claims which are not set out here. Independent legal advice should be sought from a specialist in a timely manner before pursuing a claim.

8.8 Emerging climate litigation

An increase in complaints and litigation is evident across many professional sectors, and is one of the four main drivers of 'transition risk'[74] as the UK continues its decarbonisation journey. While there is not currently a judgment in a climate litigation claim arising from subsidence damage, it could certainly be something that arises in the future. This could be a claim brought by a party outside of the insurance claim who wishes to hold either the insurance company or the tree Owner/controller to account in terms of their management of the claim and their environmental targets. This could result in greenwashing litigation or judicial review litigation where a public sector body is involved.

9 Subterranean, substructure and superstructure repairs

9.1 Introduction

9.1.1 Scope

This chapter deals with all aspects of repair including peripheral structural influences such as drains and sewers. An individual building is a single composite unit founded on a substratum. However, for the purposes of repair, it can be conveniently subdivided into three basic elements:

1. Subterranean — drains, sewers, services
2. Substructure — foundations, cellars, basements
3. Superstructure — the fabric of the building

Methods of repair are discussed and advice is given on what sort of rectification processes to use and, importantly, where to use them. Note that all work in relation to large trees and shrubs is discussed in Chapter 8.

9.1.2 Options for remedial works

When a building has suffered movement or damage, there are three options that may be considered separately or in combination. They should be approached in the following sequence:

1. Establish that the property is basically stable. Properties experience minor structural movements in their fabric from time to time, the most common being the seasonal shrinking and swelling of timber elements such as floor joists and timber partitions. This can often result in minor cracks in ceilings, wall plaster, etc. If the property is basically stable then it can be agreed that further repair to the superstructure of the building is required. These may be purely cosmetic repairs, such as filling of cracks followed by decoration. If it is thought that the movement has affected the structural integrity of the building, action may involve structural repairs, such as bonding of brickwork. The cost of repairing such damage will normally only be met by an Insurer if it is agreed that the damage arises from an insured risk.
2. If, after investigation by a suitably experienced Expert, it is thought likely that progressive movement of the foundations could continue, the next step should be to consider ground stabilisation to prevent further deterioration of the ground on which the building is standing. This could include such action as the repair of drains, pruning or removal of adjoining trees etc., after which superstructure repairs would be undertaken as previously described.
3. The last and most extensive and expensive option, if the previous two options are considered unlikely to be successful, will probably be direct remedial works to the substructure of the building, e.g., by underpinning its foundations. If a solution of underpinning is to be considered, the Insured should be made fully aware of the disruption inevitably caused and the possibility of having to vacate the property for the duration of the works.

Figure 1.3 illustrates the recommended sequence for assessing a building displaying damage and, where necessary, making recommendations for remedial works. However, if seasonal movements exceed the acceptable structural and serviceability threshold (Table 6.1) and result in major damage, special consideration may be required.

9.1.3 Expectations of repair works

It is often the case that people have different expectations of the standard of remedial works depending on their personal association with those works. Remedial works to commercial and industrial premises will frequently be accepted at standards which are below those acceptable in a domestic property. This can pose problems for those undertaking remedial works, which are of their very nature a repair exercise. This is not to excuse poor workmanship, but it is important that the parties to remedial works contracts fully appreciate and agree the standards of work to be achieved.

Experience has shown that the standard and extent of remedial works to be expected under an insurance policy claim can be a basis for disagreement between Owner and Insurer. Some Owners are pleasantly surprised at the extent of their entitlement under their policy, while others expect their property to be virtually rebuilt at the Insurer's expense, sometimes to a standard far above that which the property enjoyed before the damage occurred. It is also not unknown for Owners to claim for damage that has clearly not been caused by subsidence or any other insured peril.

It is therefore important that the scope of the work and its standard is determined and understood as far as possible between Insured and Insurers (or their representatives) before work commences and preferably even before estimates for the work are obtained.

The smooth running of a remedial works contract, such as the repair of subsidence-damaged buildings, is very much dependent on those responsible for and involved in the pre-contract investigation, administration, design, supervision and execution of the works. Once the Owner, Expert, Contractor and Insurer have agreed the extent and standard of the works and a programme for its implementation, these should only be varied in exceptional circumstances, and then only on the express instruction of the Expert administering the contract. All parties should be made aware of the reasons for any variation. Note, that variations inevitably increase costs and time, and can be a source of dispute.

9.2 Ground stabilisation

9.2.1 Scope

This course of action is best considered as remedial work or modifications to features in the vicinity of the property but which do not form an integral part of the building structure. Such items would be, for example, attention to major vegetation and trees within influencing distance of the building, or drains and sewers which are sufficiently close so that leakage from or into them could affect the ground in the vicinity. The presence of subterranean streams and old filled ponds can also be included in this category. Typical examples of corrective action are provided in the following three subsections.

9.2.2 Repair of drains and leaking water mains

The replacement of worn out, badly maintained or poorly constructed drains is not a matter for an insurance policy. However, accidental damage to drains is often insured. An Insurer would wish to establish that the reason for the damage of the pipes is indeed accidental in origin, typically where pipes have been affected by an unforeseen external factor. This may include damage by tree roots, trafficking over shallow drains or perhaps even general ground movement. Damage to shallow drains may be the result of poor drainage design.

Drains may also be damaged by additional work to the original structure, for example, the construction of an extension or even the erection of a fence or garden wall.

Where damage by subsidence to a building occurs as a direct result of leaking drains, some Insurers may consider the damage to the building under the 'Escape of Water' section of the policy. Insurers are principally concerned with the initial or proximate cause of damage. Where drainage repairs are required these should be actioned as quickly as possible. Where the cost of drainage repairs is not covered by an insurance policy, yet drains are allowed to continue leaking, there may be a detrimental effect on the foundations of the building. Early action is therefore essential to mitigate loss. Subsequent damage may not be covered if wilful neglect of maintenance is the principal reason for the damage.

Where soil rehydrates as a result of other causes — typically leaking gutters and path-shedding of water towards the property — the cost of rectifying these defects is not normally a matter for the Insurer, although if subsidence damage occurs as a result of these other causes, a valid claim may still arise unless there has been wilful neglect on the part of the Owner.

Lack of attention to drains can cause subsidence by permitting water to escape into the ground, softening the substratum beneath the foundations. Once softened, it is possible that the ground may not be capable of supporting the load and so the building might subside in the vicinity of the leaking drain (Section 6.4.2.6). Leaking drains can attract tree roots to their vicinity, often close to foundations, and assist sulphate ingress if appropriate materials are present in the ground. Alternatively, a defective drain may permit water from the ground to leak into it and this might wash away the ground from beneath adjacent foundations, causing subsidence.

Repair of the drains, or laying new drains and sealing the old ones should prevent further leakage into or from the ground. However, it may still mean that some subsidence could continue for a year or so after the remedial works have been carried out because the ground will need to dry out and consolidate where it has been softened. If the ground beneath the foundation has been washed away, merely repairing or replacing the drain may not be sufficient to prevent movement in the future, and pressure grouting of the ground or underpinning may be required.

Leaking water mains should only be dealt with by the water authority or one of their approved contractors.

9.2.3 Control of ground moisture content

Some older properties have relatively shallow foundations and as a consequence the substratum can dry out simply by evaporation of moisture from the ground during particularly dry summers — without the involvement of trees or other vegetation. If this evaporation causes shrinkage of the substratum, as can be the case with some types of clay, then subsidence may occur (Section 5.2.3).

Concrete flags or *in situ* concrete paths of 1.5m in width around a property which has foundations less than 1m in depth is frequently sufficient to cure all but the most extreme cases. Likewise, removal of surrounding paths or adjacent buildings can suddenly encourage seasonal movement by permitting drying and wetting of the ground on a seasonal basis — in ground which was previously relatively dry but stable.

Additionally, the 'rain shadow' effect caused by lightweight canopies or overhanging dense vegetation can lead to excessive drying of soils where rainfall fails to reach the ground.

9.2.4 Soil moisture management

The removal of moisture from the ground by tree roots is by far the most common cause of subsidence damage to buildings. As has been referred to previously, this occurs most frequently to properties constructed on shrinkable clay soils. The ability to be able to control the moisture content of the ground despite the activities of the trees in the vicinity would clearly have substantial benefits. Experiments have been undertaken in an attempt to develop systems for controlling, or more precisely managing, the moisture content of the ground.

To date no method of rehydration has been proven to be commercially satisfactory. However, research into methods of rehydrating soils is continuing.

The changes in seasonal moisture content of fissured clay soils can be mitigated by the injection of specialist expansive geopolymers. The objective of such treatment is to fill in the natural fissures and discontinuities within the clay with a low permeability geopolymer. By doing so, the bulk permeability of the fissured clay will typically be reduced by a factor of 50 according to Buzzi et al[75].

9.3 Remedial works to foundations

9.3.1 Scope

This section applies to work carried out on the structure of the building or the ground immediately beneath it, such as:

- Providing new forms of foundations — normally termed 'underpinning'
- Improving the quality of the ground beneath the foundations by injecting cementitious grout or geopolymers into the ground — normally termed 'pressure grouting'
- Jacking the building back into position — a procedure which requires very careful assessment

In the 1980s and 1990s, tens of thousands of domestic properties were underpinned throughout Britain. As a result, many techniques for underpinning have been developed to a high degree of efficiency by specialist Engineers and Contractors.

Imperative to all underpinning works is the necessity to have a thorough appreciation of the ground conditions beneath the property in question. The quality of the ground will almost certainly have been the primary cause of the subsidence. An adequate site investigation to determine the character of the substratum is essential to establish:

- The depth and nature of the poor quality ground and of the ground to be used for support
- The depth and type of existing foundations to the property

- The presence or otherwise of groundwater — this can affect the works and can also alter the ground characteristics during and even long after the remedial work has been undertaken
- Whether there would be any advantage in undertaking remedial works at a particular time of year to account for seasonal variations

The investigation should also note the location and type of below ground services in the area which could interfere with remedial works.

Any surrounding vegetation and trees should be taken into account. During the ground investigations, roots in, around or below the foundations should be logged so that the design engineer can allow for these. This may also be important if action is to be taken against a neighbour.

The type and extent of the site investigation will almost certainly depend on the individual circumstances. In most instances of minor subsidence, a simple series of trial pits (to determine the existing footing depth and detail) with some hand augering to a depth of about 3m will suffice, subject to the need to establish the depth of the stable stratum.

In other cases the investigation will need to be much more sophisticated and extensive, particularly if piling techniques are envisaged. Site investigation techniques and their suitability for different circumstances are discussed in Chapter 7. No remedial ground works should be contemplated without a site investigation being carried out beforehand. A suitably experienced Engineer is the most qualified Expert to advise on the type and extent of the investigation required and therefore on the most appropriate underpinning system for the particular substratum conditions.

The most commonly used underpinning systems are described in Sections 9.3.2–9.3.4.

9.3.2 Traditional mass concrete underpinning

Mass concrete underpinning is very commonly used in instances where the total depth of excavation does not exceed 2.5m. Beyond this depth, careful consideration by the Expert as to the risk involved of placing operatives below ground to this level must be made. Under CDM 2015[15], the design process must consider ways to design out risk and the Expert must consider if safer, more economic methods with less environmental impact exist. For greater depths, other forms of pile-based underpinning tend to be less hazardous, more economical to install and have less environmental impact than mass concrete underpinning.

The method involves the insertion of individual blocks of concrete beneath an existing foundation (Figures 9.1 and 9.2). Such blocks normally form a continuous footing beneath the existing foundation. They are inserted in lengths of approx. 1m and in a sequence so that no more than one third of the existing foundation is unsupported at any time. Sometimes even more conservative spacing of blocks is advisable.

Figure 9.1: Mass concrete underpinning in progress

Figure 9.2: Mass concrete underpinning shuttered and cast against adjacent pin

Alternatively, blocks may be inserted at isolated locations around a property with the original ground remaining in place between the blocks, and the existing foundations spanning across between the individual blocks (Figure 9.3). The preferred arrangement is for the blocks to form a continuous footing because that way any tendency for slight differential movement to occur between the blocks is virtually eliminated. The blocks should be interlinked to maintain continuity between adjacent blocks and this can be achieved by toothing out to form a 'joggle joint' or 'keyway' for the adjacent block to connect into the adjacent block (Figure 9.4).

Figure 9.3: Isolated block mass concrete underpinning

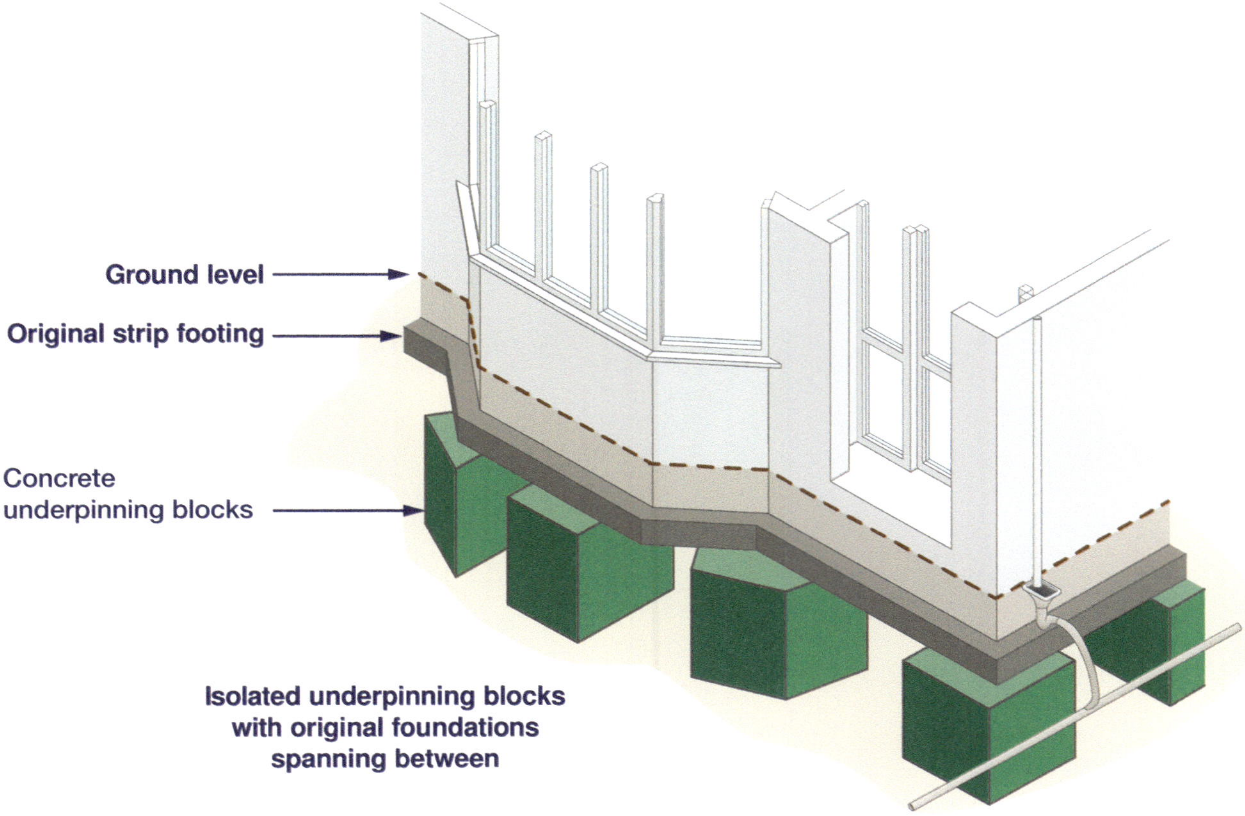

Figure 9.4: Continuous mass concrete underpinning

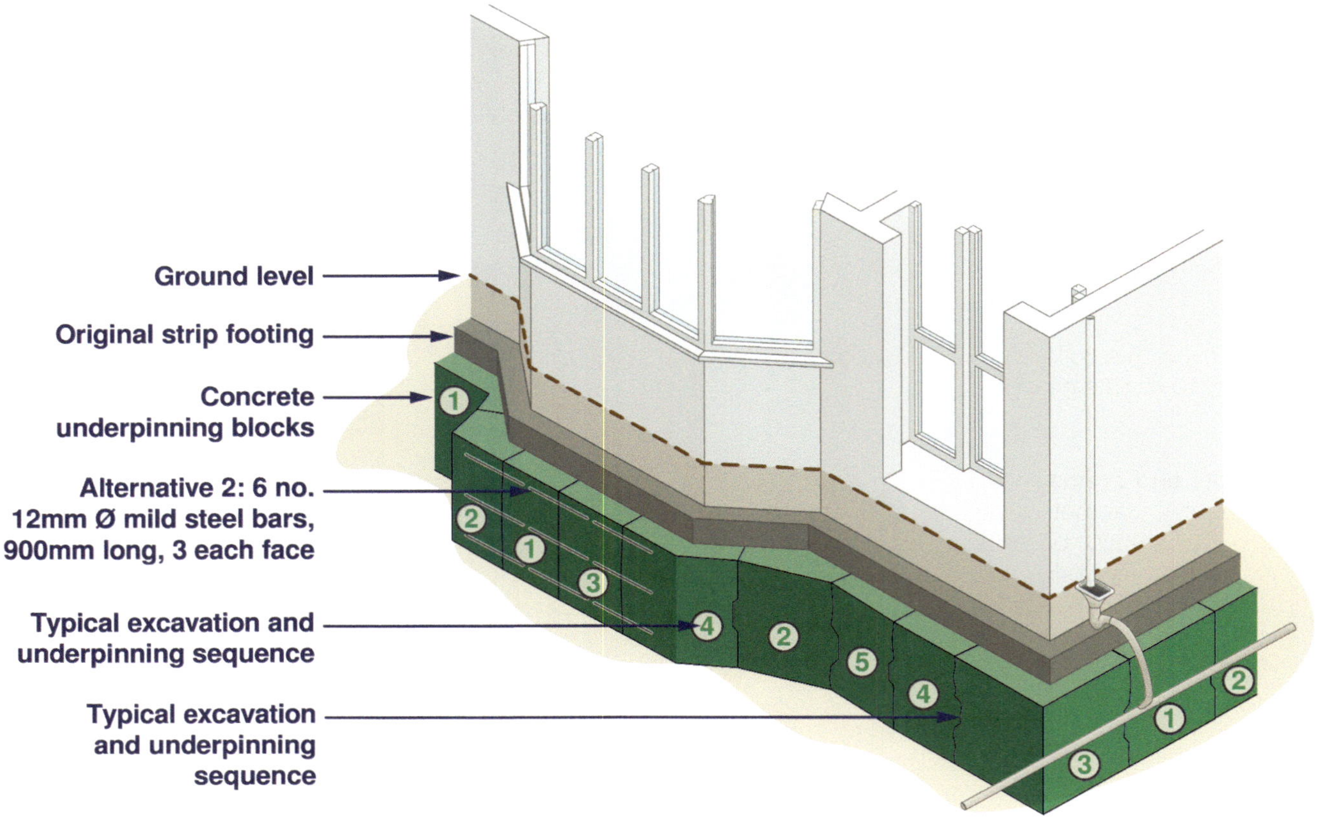

The sequence of excavation and casting of the blocks by either of these methods should provide adequate support at corners and points of load concentrations, particularly during temporary conditions.

9.3.3 Beam & Base underpinning

This is similar to the isolated block system described in Section 9.3.2 but it incorporates a structural beam (Figure 9.5) spanning between individual blocks to support the intermediate section of the original foundation. As such, the original foundation does not have to span between the blocks of its own accord. A series of deep square concrete bases

Figure 9.5: Site ready for beam to be cast

(pads) are normally excavated and formed at intervals beneath the walls of the damaged property. The beams are then formed or placed under the intervening parts of the walls to span between the bases, i.e., supporting the walls (Figure 9.6). Both this and the previous system require considerable excavation and disposal of ground material. This can sometimes adversely influence the use of these types of underpinning procedures.

Figure 9.6: Beam & Base system

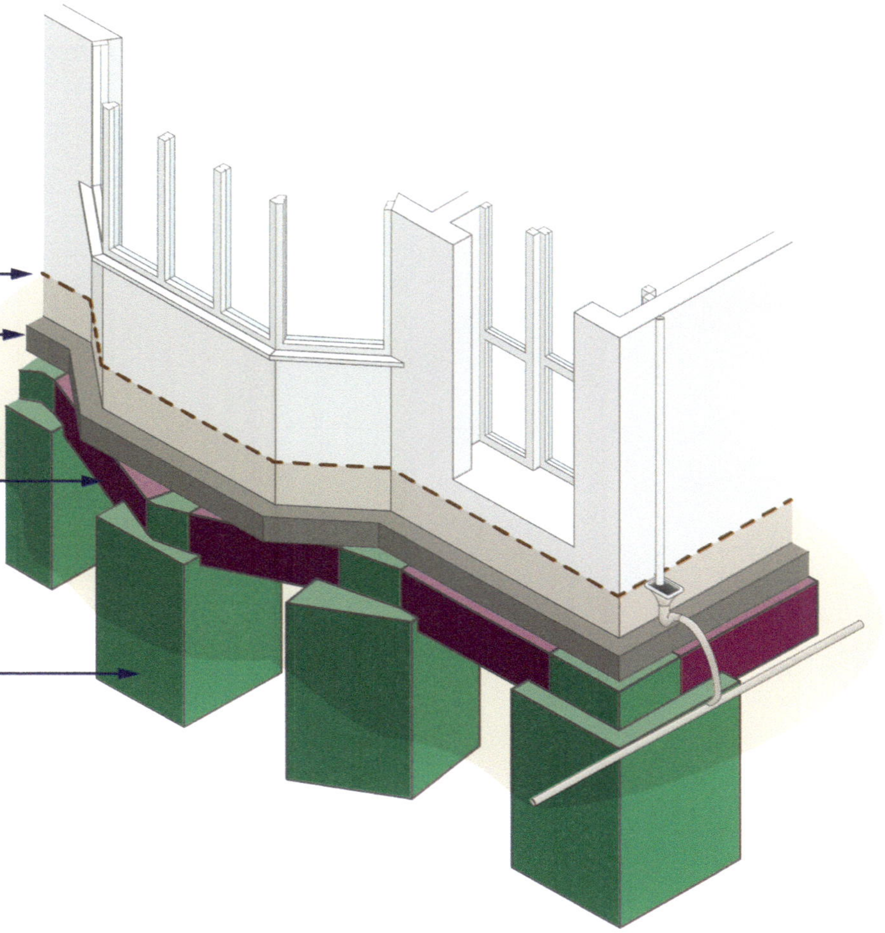

When designing and installing Beam & Base systems, it is important that the following points are taken into account:

- The possibility of roots from adjacent trees growing below the beams and effecting the internal foundations in the future
- The likelihood that the excavation for the underpinning and the backfilling around it may provide a pathway by which groundwater can percolate to the base of the underpinning, causing softening of the substratum

9.3.4 Mini-pile underpinned systems

Piling consists of the installation of a steel, concrete or composite shaft through weak ground (by driving or boring) that is capable of transferring the forces from a structure into or onto a suitable bearing strata. A mini pile is simply a small diameter pile but is designed using the same design standards and engineering principles found in BS EN 1997-1[76] and BS 8004[77].

Most mini-piling systems used as part of a subsidence repair system are designed to take only direct compression loads. Tension, shear and bending can be accommodated; however, in order to complete such a design it may be necessary to undertake additional site investigation which would include soil strength parameters and engineering descriptions.

The main types of mini-pile and their method of installation are:

- **Driven displacement piles.** An internal drop hammer weight (Figure 9.7) or a vibrating pneumatic mole is used to drive steel casings to a predetermined 'set'. The hammer drives down through the centre of the tube against a concrete plug at the toe of the casing. Casings are typically in 2m lengths and as the pile progresses into the ground additional sections are added and welded together. Once in a set or at design length, the pile is filled with reinforced concrete which becomes the structural element. Generally used in filled ground, granular soils or where water in encountered — as small driven piles commonly struggle to achieve adequate design length in clay soils.

Figure 9.7: External piling with mini-rig

- **SFA (Sectional Flight Auger) piles.** Hydraulic rotary mini-piling rigs are used to auger using either solid stem flights by themselves in stable ground, or solid stem flights with casings (or segmented hollow stem augers) in unstable ground (Figure 9.8). Once at design length, the piles are filled with either concrete or grout from the top, tremmied from the base, or in the case of hollow stem augers, through the centre of the auger as the flights are extracted. Generally used where cohesive soils exist — particularly clay, even if unstable ground exists at higher levels.
- **Drilled piles.** Commonly used through obstructed ground or where penetration into rock or very dense ground is required. Often used on projects associated with mining subsidence. Typically, a pneumatic tungsten tipped auger is drilled into the ground using a rotary piling rig and a string of drill rods. Often, in unstable ground the drill will take a steel casing into the ground to support the pile shaft, and once the design length has been reached the drill is withdrawn and the pile filled with reinforced concrete.

Figure 9.8: Hydraulic rotary mini-piling rig

- **Jacked piles.** Precisely controlled hydraulic jack sets (Figure 9.9) are used to jack mini-piles into the ground. A resistance has to be created against which to jack, which often takes the form of holding down bolts connecting the rig to a concrete raft under the building (either new or existing). All jacked mini-piles are proof loaded by virtue of the installation method, though careful monitoring during installation is needed to prevent issues (Figure 9.10). The major benefit of this system is that it is totally vibrationless which is a major advantage when dealing with distressed structures. Can be used in most ground conditions. Preformed mini-piles such as 150mm square precast or steel sections are most commonly used.
- **Screw piles.** Steel hollow sections with a series of helical plates along the shaft are screwed into the ground to either a design length or to a predetermined torque. Several variants of this pile type are available on the market and can be installed using rotary piling plant, modified excavators or pneumatic driving hammers.

Figure 9.9: Jacked piles

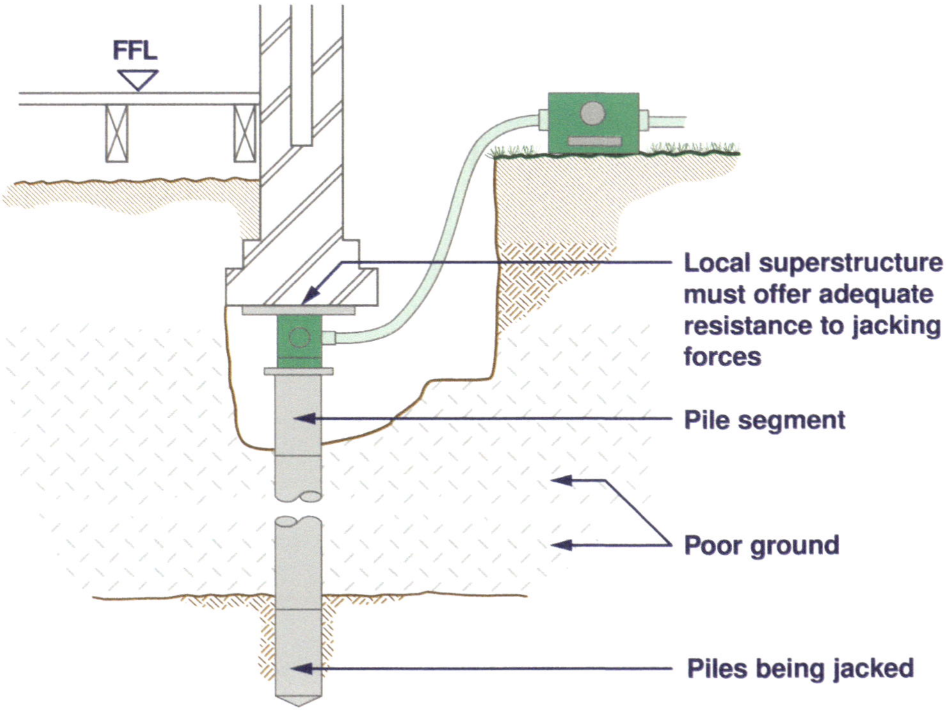

Figure 9.10: Possible problems during installation of piles

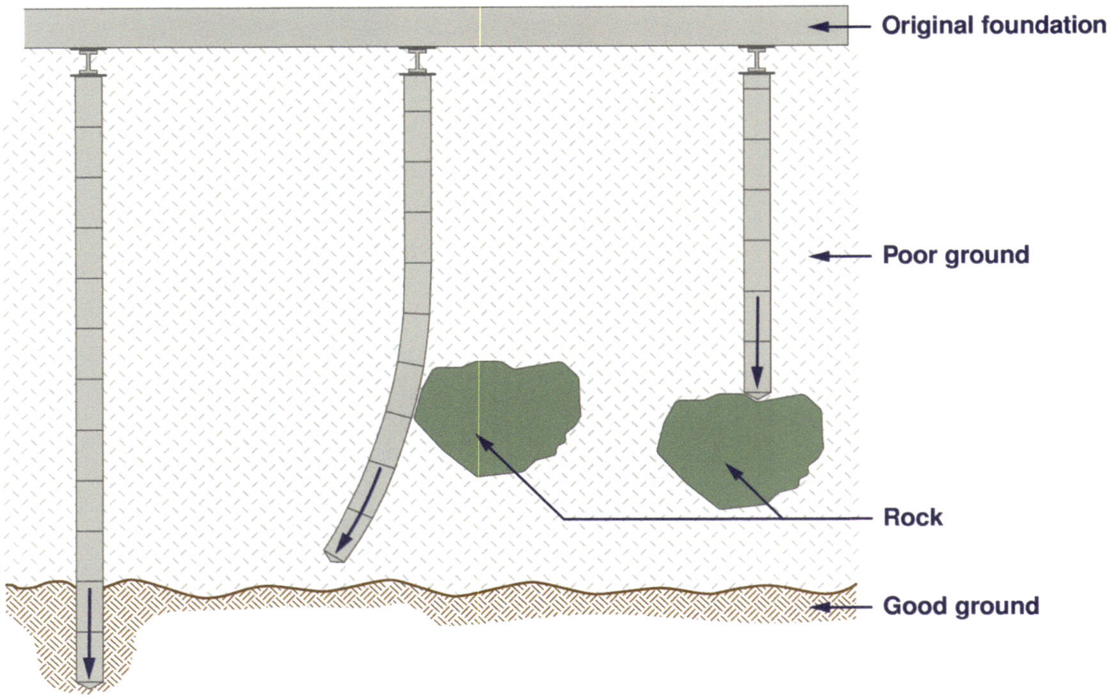

Where anti-heave precautions are required, the piles must be designed to resist heave forces from the shrinkage and swelling of the ground. Agreement should be sought from the design team as to the potential depth of tree influence. This should be used to calculate the tension and compression heave forces on the pile, and the design engineer must then satisfy themselves that the pile has sufficient embedment and capacity at depth to resist these forces. The pile shaft must also have sufficient structural capacity to transfer these forces within the pile section.

There are products on the market which may help to reduce heave forces such as heave sleeves and slip liners which claim to reduce the coefficient of friction between the soil and the pile — but their use and benefit is not universally accepted and good quality site control is required especially when using grouted piles.

- **Scissor or root piles (also known as angle or stitch piles).** These are piles, usually small in diameter, drilled or driven through the foundation of the property to support the property directly by forming a connection between the existing structure and the top of the pile (Figure 9.11). Piles inserted by this procedure are raked at an angle or installed in vertical and raked pairs. The procedure has been in practice for many years and works very successfully, not only for low-rise buildings but also for substantial structures. The procedure does rely greatly on the workmanship associated with the installation of the piles and the condition and size of the existing substructure for its successful implementation.

  The design of this type of system needs to take diligent account of the existing foundation geometry, so as not to over stress the piles by inadvertently transferring high shear and bending moments into slender pile sections. Ways of overcoming these issues include using pairs of piles, or by connecting internally-raking piles to concrete floor slabs.

- **Pile & Beam installations.** In this procedure piles are installed vertically on both sides of the wall to be underpinned and a reinforced concrete or concrete encased steel beam is then positioned spanning between two or more piles to support the weight of the existing wall (Figure 9.12a). The beam may pass through the existing wall (Figure 9.12b) or be situated immediately below the existing foundations (Figure 9.12c). The system can consist of a series of individual pairs of piles with a beam between, or a zig-zag pattern consisting of a number of interlinked piles and beams (Figure 9.13). This is a very flexible and well-proven procedure. It requires a relatively small amount of excavation, but does require specialist design and installation.

Figure 9.11: Scissor or root piles

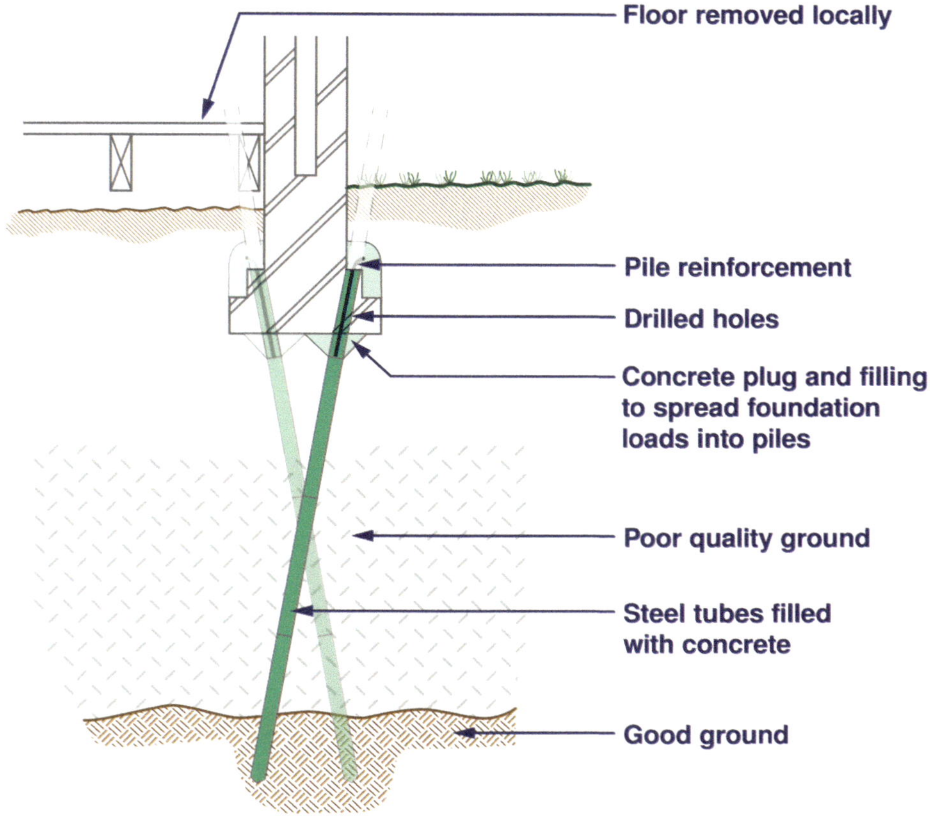

Figure 9.12: Pile and beam installations showing alternative positions of ground beams: (a) plan view; (b) above footing, (c) below footing

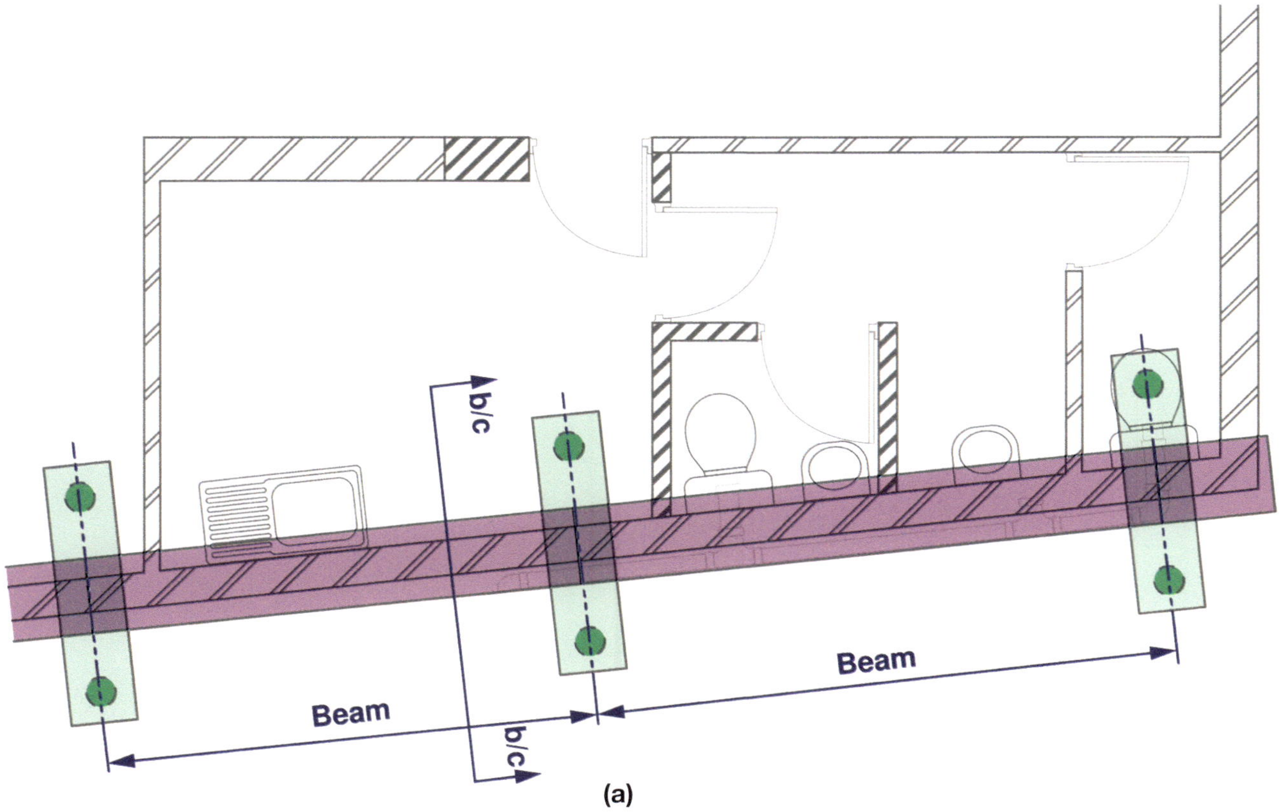

(a)

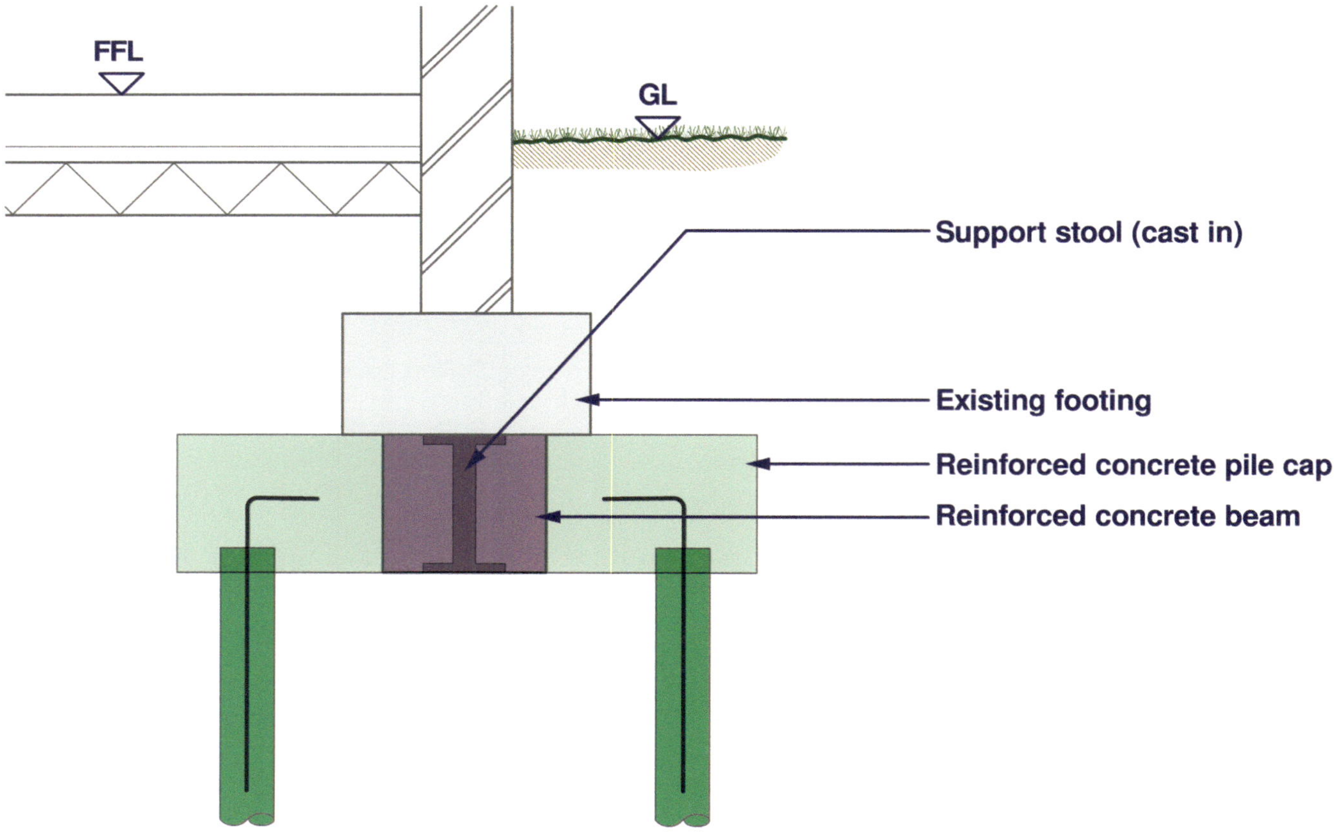
FFL
GL
Support stool (cast in)
Existing footing
Reinforced concrete pile cap
Reinforced concrete beam

(b)

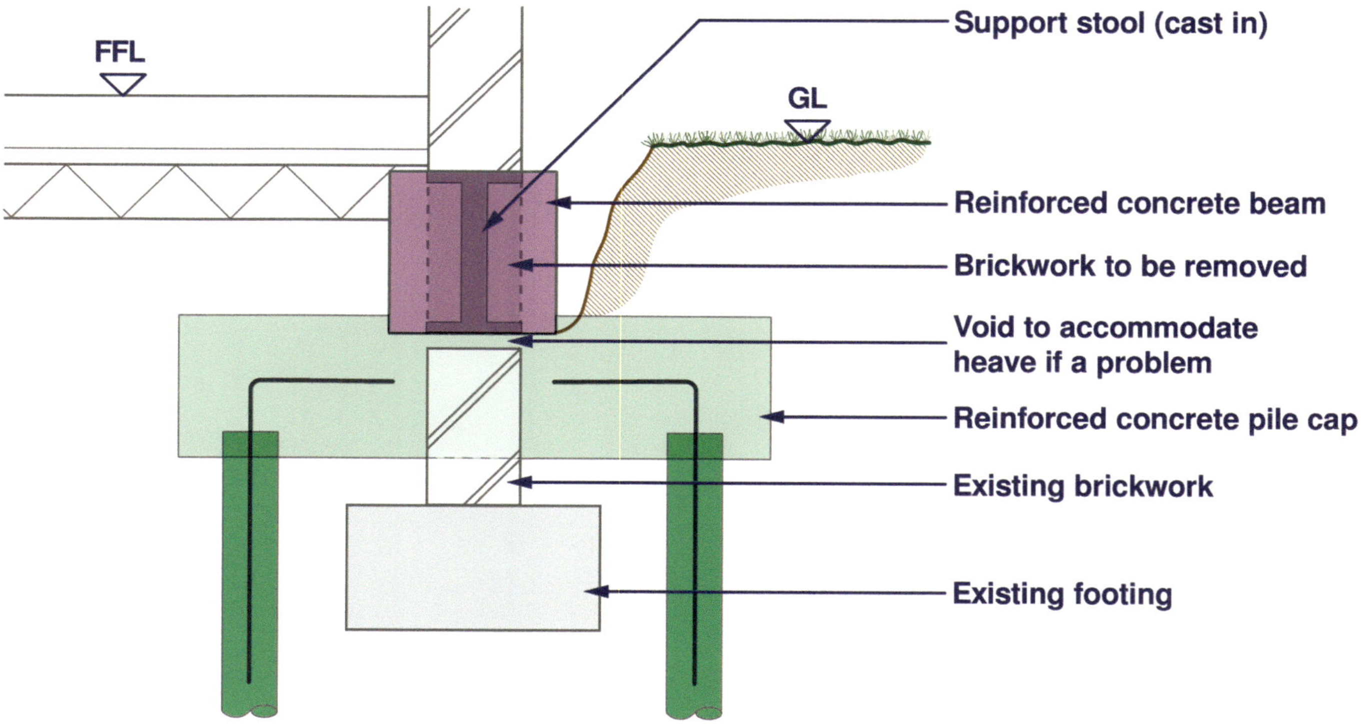
FFL
GL
Support stool (cast in)
Reinforced concrete beam
Brickwork to be removed
Void to accommodate
heave if a problem
Reinforced concrete pile cap
Existing brickwork
Existing footing

(c)

Figure 9.13: Zig-zag arrangement of pile and beam

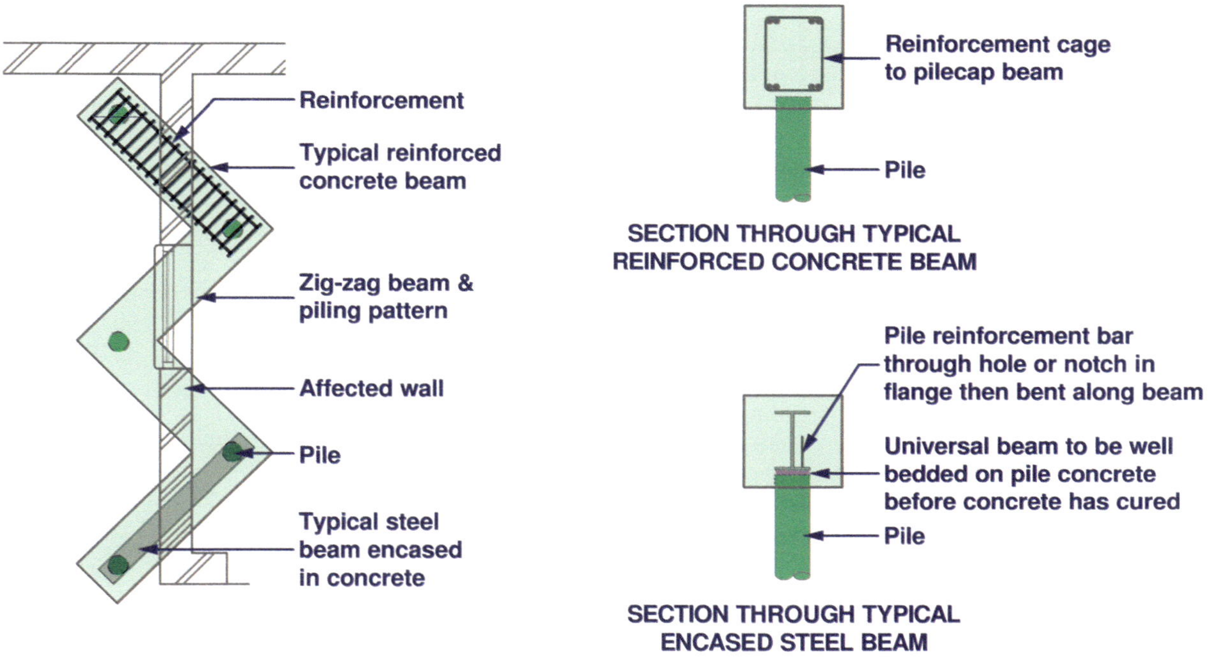

- **Cantilevered beams.** It is possible for a cantilever type of arrangement to be constructed by installing the piles on one side of the wall only and permitting the beam to pick up the weight of the wall on an overhanging cantilever (Figure 9.14). This cantilever arrangement relies on the pile closest to the wall being in compression, and the one furthest from the wall in tension. Care needs to be taken by the designer to ensure the ground conditions and length of pile shaft are adequate to resist the tension forces – if not this can be counterbalanced using kentledge or an oversized pile cap.

Figure 9.14: Cantilevered beam

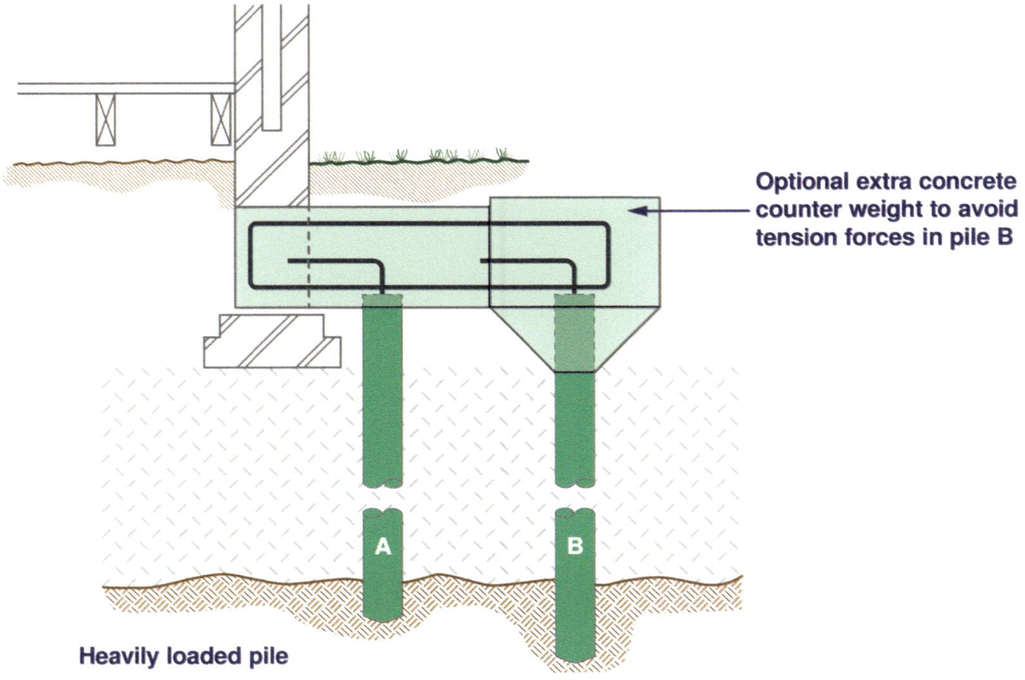

The system has the advantage that it can be installed from one side of the affected wall only, and both beams and slabs can be used to transfer the wall loads into the piles.

- **Single piles with cantilever or bracketed heads.** This system has been specifically developed to be used in confined or restricted access situations. As with all mini-pile solutions, care must be taken to ensure that design and installation is adequate and the workmanship is well supervised.

  This system is used where relatively lightly loaded walls are to be carried and a single pile is theoretically adequate to support the vertical load. The pile is installed on one side of the wall and a bracket of steel or reinforced concrete is affixed to the top of the pile and into the adjacent wall. This system in its basic form will result in a bending moment being transferred to the pile head due to the eccentric vertical load. This can be resolved in the pile section, provided the near-surface soils have sufficient stiffness and the pile section is reinforced to resist the resultant bending moment. Alternatively, the wall requires a horizontal restraining force close to the line of the brackets to stop the piles bending. Where bracket piles are installed either side of a wall in an alternate sequence the stability is improved, but tortional forces can be transferred into the existing substructure.

- **Rafts supported on piles.** The principle of cantilever beams can be extended with greater certainty to a raft type system supported on piles. This arrangement uses the counterbalancing effect of two walls to transfer the load of one or both walls onto the piles. This is a more predictable structural arrangement that tends to eliminate any undesirable tensile forces in the piles. The piles are subject to compressive forces, which they can accommodate much more easily. The raft and pile arrangement includes a toothing system of reinforced concrete nibs which are cast into holes cut in the walls to be supported, or in the walls to be used as counterweights (Figures 9.15–9.17).

  The system is straightforward although care should be taken to ensure that the toothing arrangement is properly constructed, so that the walls are picked up adequately by the reinforced concrete nibs which are cast into the wall. Compaction of the concrete to the nibs can be an issue, so a suitable mix design should be selected to allow adequate compaction and ensure load transfer from the walls to the new RC structure. The property must then be isolated from the existing foundations to prevent any heave influence.

Figure 9.15: Reinforcement being placed ready to form raft

Figure 9.16: Raft ready for concrete pour

Figure 9.17: Raft supported on piles

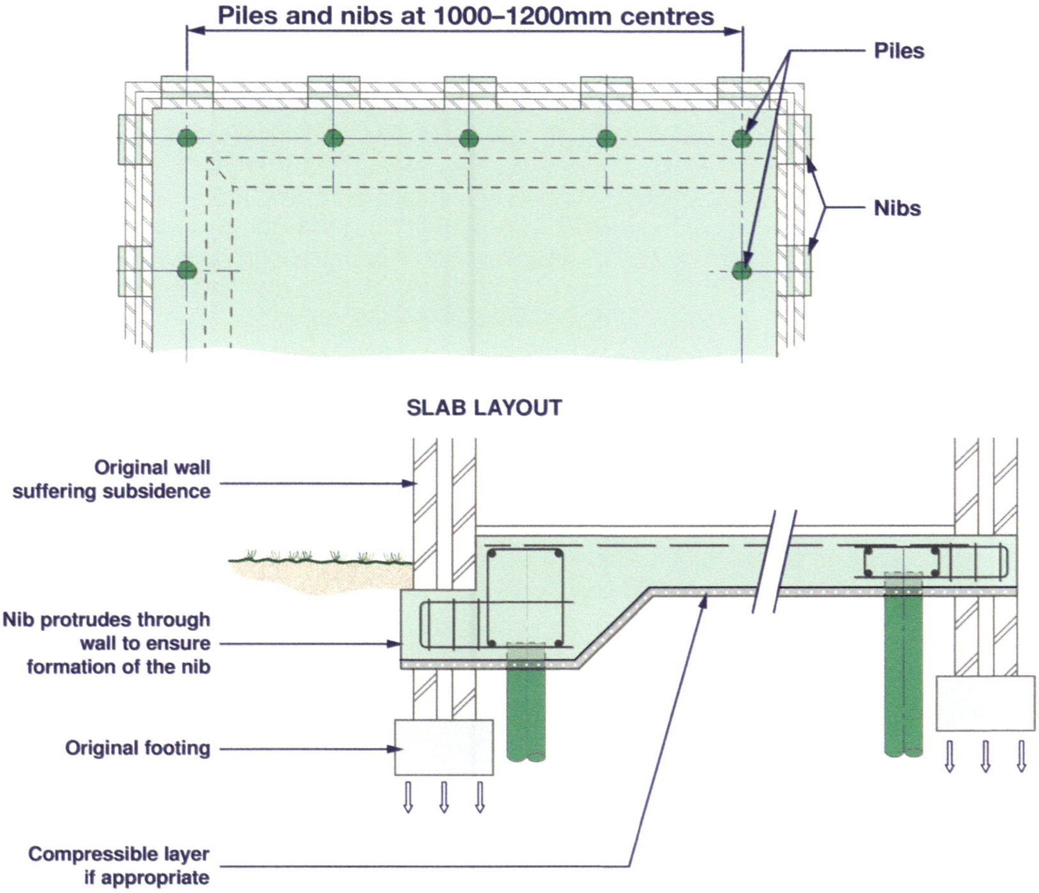

- **Jacked piles.** This system is normally used where access is restricted to one side of a wall and where the part of the structure to be underpinned is of considerable dead weight and there are no openings in the lower part of the wall. Jacked piles are less common than other piling solutions, but do have their uses where the ground conditions are predictable and consistent, and the depth to which the piles will be inserted can be established with reasonable certainty.

Unlike most piling systems, the installation of jacked piles does not normally cause vibration or noise, and in certain circumstances this can be very beneficial. However, this system is only suitable where sufficient weight exists in the part of the structure directly above the hydraulic pressure jack to accommodate the jacking load. If the wall does not have sufficient weight, it is probable that the wall itself and anything which bears onto it will be pushed up out of the ground rather than the pile being pushed into the ground. One way to overcome this is in the design of the jack piled system, and cantilevered edge beams will allow mobilisation of longer sections of wall — but with this solution more heavily reinforced sections are inevitable.

Jacking also imposes locally concentrated forces on the building at the jacking positions. It is important that the construction directly above the jack is able to withstand the crushing effects of the jacking load. If there is doubt on this point, bearing plates must be provided. Care must be taken to avoid local damage. Alternatively, an adequate temporary load (kentledge) should be provided against which to jack the piles.

Successive lengths of pile are added on top of preceding lengths until it is impossible to jack the piles down any further for a given load on the jacking system. The advantage of this system is that the load on the pile can be positively established at the time the pile is installed.

Considerable skill is required in the installation of piles by this system and it should only be undertaken by Contractors who are competent and have specialist knowledge of pile fabrication and jacking procedures. If piles go appreciably out of plumb during jacking, they may have to be abandoned and new piles installed or a completely different technique adopted. Monitoring of the installation procedure for jacked piles is very important, particularly to check that the load on the pile is the genuine load and is not influenced by factors which could produce misleading readings.

- **Pressure grouting.** Although strictly speaking pressure grouting is a ground stabilisation procedure, for the purposes of this guidance it is included here as a remedial works system. Pressure grouting is a procedure by which poor quality ground is improved by injecting a cementitious mixture into the ground to improve the bearing capacity of the substratum. Pressure grouting is most suitable for ground which contains voids or which consists largely of granular type material, such as gravel, rock fragments, brick hardcore fill or similar.

  Little if any excavation is required. The grout is injected into the ground on a predetermined pattern. Sometimes a grout barrier is first created in the ground around the property and then the ground within the barrier treated, while on other occasions grout is inserted over the area involved on a simple grid pattern. The procedure is determined by:

  - The ground conditions
  - The type of building being treated
  - The accessibility onto the site for the grouting machinery

  Old mines, caves and similar ground hazards are frequently treated using pressure grouting procedures. Pressure grouting is not normally suitable for substrata containing clay unless the problem is the formation of voids within the clay.

  A major problem is that it is sometimes difficult to predict precisely where the grout will disperse within the ground. As such, some Insurers may be reluctant to accept this as a permanent solution. On occasions, adjacent cellars or drains have been grouted inadvertently and even cavity wall voids above ground level have been known to be filled by mistake.

  Pressure grouting procedures for domestic underpinning have been used successfully where the nature of the substratum makes it appropriate. Pre-works estimates of the amount of grout likely to be used are an important safety measure.

9.3.5 Jacking building back into previous position

It is occasionally possible to jack part or all of a building back to its previous level or position, but this procedure must be done with great care so that the part of the building in question is not further damaged during the operation. It is important that the section of the building to be jacked back must be absolutely free to move, and sufficiently rigid within itself to avoid suffering damage from any slight discrepancies which may occur in the jacking procedure. Small sections of a building, such as single storey extensions founded on rafts, can often be treated without much difficulty, but treating large sections of structure by this procedure can be hazardous. This is always a specialist operation.

Before the jacking process commences, a new foundation from which the jacking operation can be carried out, must be inserted below the existing foundations. Jacking a building back into position should only be contemplated if its distortion is significant and unacceptable. The Expert should be satisfied that the jacking procedure will be cheaper and safer than taking down and reconstructing the affected part of the building. Services into the building which may not have been damaged by the original subsidence, or which may have been repaired to accommodate the original movement, are likely to require further repair after the jacking operation has been completed.

9.3.6 Geopolymer injection

Geopolymer injection is a non-disruptive ground improvement technology which is installed via small diameter (14–16mm) steel injection tubes, which are drilled into the ground using lightweight equipment. The geopolymers are multi chemical liquid component systems which are thoroughly mixed at the point of injection into the ground. This triggers an exothermic chemical reaction, which induces a state change from liquid to solid. It is the consequential increase in volume which produces dynamic energy and subsequent expansive force. The chemical reaction typically takes less than 60 seconds and results in a solid compound, which possesses definitive physical and mechanical characteristics. The hardened geopolymer reaches 90% of its final mechanical strength within 15 minutes and 100% within 24 hours. The expansion force created by the geopolymer injection compacts and densifies granular materials and strengthens clay soils by filling existing and newly created micro fractures with hardened geopolymer.

The geopolymer injection design varies according to the desired objective which can include:

- Reduction of shrink-swell behaviour in clay soils
- Increasing bearing capacity and reducing settlement
- Lifting floor slabs
- Lifting buildings
- Filling voids
- Permeability reduction

Geopolymer injection design and execution should be undertaken by expert specialist contractors. A typical geopolymer injection profile below a conventional strip foundation is shown in Figure 9.18 and a plan view of injection points is shown in Figure 9.19. Pictures showing the typical placement of injection tubes and the works progressing are shown in Figures 9.20 and 9.21 respectively.

Figure 9.18: Typical geopolymer injection (profile with depth)

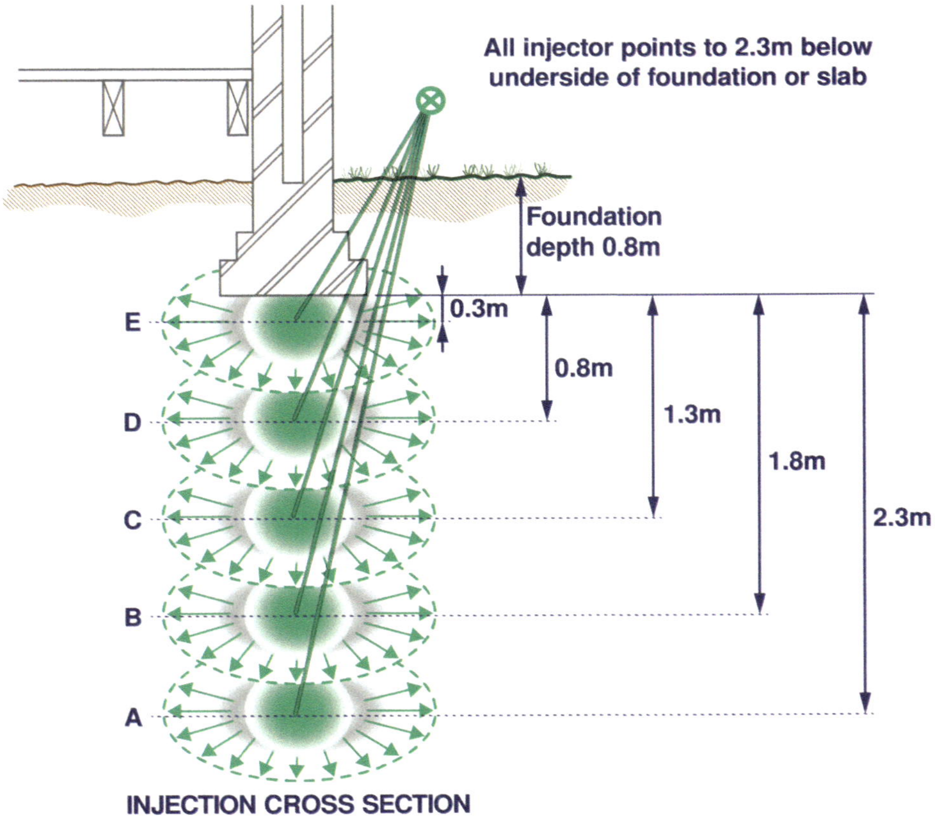

Figure 9.19: Typical geopolymer injection (plan view)

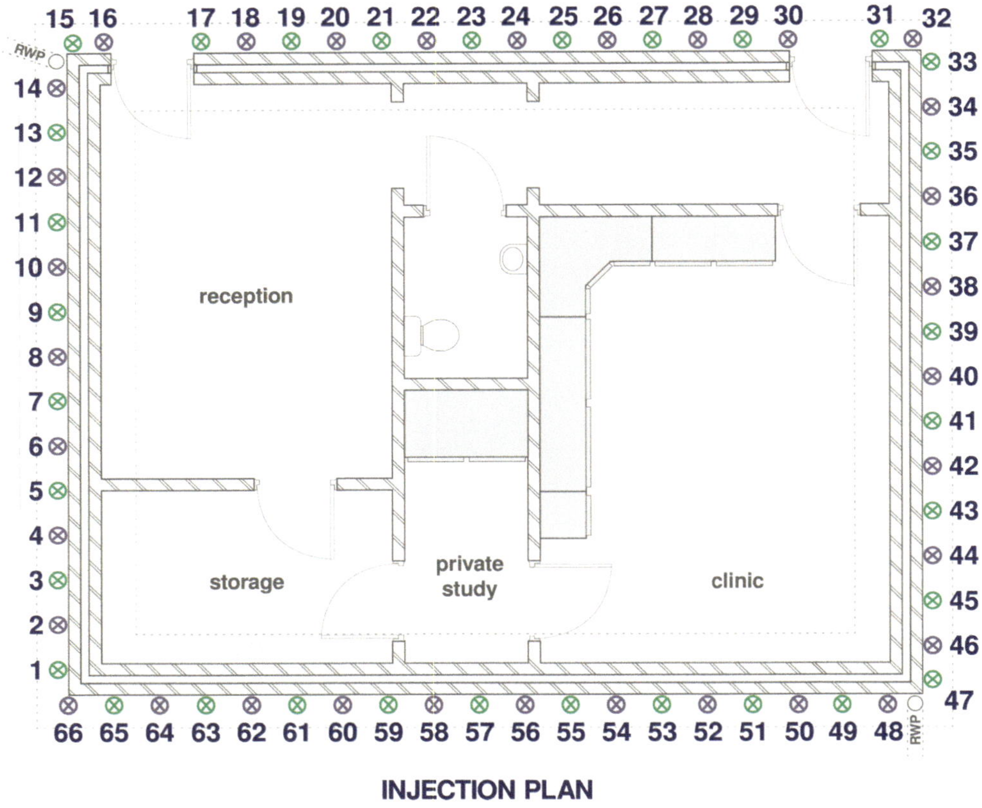

Figure 9.20: Geopolymer injection tubes in place

Figure 9.21: Geopolymer installation works in progress

9.4 Guidance notes for those responsible for remedial and underpinning works contracts

The guidance in this section largely takes the form of lists of practical considerations when implementing remedial works contracts; they are not exclusive and each contract should be judged on its own merits. They do not contain advice on such matters as tree removal, soil stabilisation, etc. but address underpinning per se. Above ground remedial works to walls etc., is covered in further detail in Chapter 10.

Having decided that a structure requires underpinning, the Expert and contractor (when appointed) will need to assess a number of separate issues. The issues will relate to such matters as the practicalities of executing the work, the specification for those works, and to aspects of administration that perhaps do not apply to superstructure or drainage repairs.

All piling, underpinning and polymer grout injection works are skilled operations and require a fully trained, experienced, competent specialist contractor to undertake them. The use of general builders or inexperienced and untrained trades should be avoided.

Some of the issues likely to arise are listed in this section and its subsections. It is not possible to specifically say which party should normally be responsible for which issues; the Conditions of Contract for the particular job may determine some of the responsibilities. Note, that where possible, the risks and the party best placed to deal with these risks, should be identified at an early stage.

For the specific underpinning techniques described in Sections 9.4.1–9.4.5, it is assumed that the Expert will be primarily responsible for considering the risks listed. This is because it is the Expert who will eventually be issuing the Certificate of Structural Adequacy (Section 10.8 and Appendix B) but there are matters for which the contractor (when appointed) will clearly be responsible, e.g., the conduct of the contract on site and the workmanship.

General items

Note: where appropriate in this (general items) subsection, a suggestion as to the 'responsible party' has been made.

Specifying the works
Generally it is preferable for the Expert to specify fully the contractor's obligations rather than place reliance on potentially inadequate performance specifications.

Superstructure loads
The loading arrangement of the building should be closely examined so that the loads on the underpinning works are correctly assessed and catered for in design.

Ground conditions
Will the underpinning be carried out traditionally? Is there standing water? Does the initial appraisal (site investigation) provide the necessary level of information to design a solution adequately? (Primarily a matter for the Expert).

Site investigation
The site investigation should match the anticipated remedial works scheme. Basic site examination is permissible for mass concrete underpinning and similar procedures but it must still be adequate for its purpose. For piling works, the ground should be proved to at least a depth 50% greater than the anticipated length of the pile. (Primarily a matter for the Expert).

Health and Safety
As per CDM 2015, who will fulfil the roles? (client designer and contractor?). Do they both have adequate knowledge and experience? Is the project notifiable to the HSE?

Mini-piled underpinning
Will ground conditions be conducive to mini-piling? Will access be available for mini-piling equipment? (Primarily a matter for the Expert). Note: For the purpose of this guidance mini-piles are defined as piles having a diameter of 300mm or less.

Design and/or construct
Many of the trading underpinning contractors execute their own designs for which they have PI or other insurance — do the proposed contractors have this capability? Are they financially stable? Will the chosen contractor be able to deal with the complexity of the project? (Primarily a matter for the contractor).

Inspection by the Expert
The Expert should also be prepared to inspect for themselves as reasonably appropriate during the contract, but note that responsibility for achieving the specified quality of workmanship always rests with the contractor.

Building control inspections
The works fall within the scope of Building Regulations (primarily a matter for the Expert). Have the necessary notices been served? Is the intention to use Local Authority, or private building control?

Planning and Listed Building consent
Is the property listed and or in a conservation area? Who will check and if needed who will make the application, heritage statement etc?

Party Wall awards
Will the piling and/or underpinning works require a party wall award to be served on adjoining Owners?

9.4.1 Guidance for traditional mass concrete underpinning

The following additional risks/issues should be considered when the Expert specifies the extent of the works and the materials to be used for this type of underpinning:

- 'Ready-mix' vehicles require good access and on occasions must deposit the concrete some distance from where it needs to be placed unless site pumps are available. However, ready-mix does deliver a consistent quality of concrete, provided one of the standard mixes has been specified

- In all cases it is the responsibility of the Expert to specify the minimum period before new foundations can accept the imposed load of the existing building
- Load transfer from the structure to the underpinning is achieved by one of the following methods:
    - o Dry packing (Figure 9.22a). The underpinning concrete is left 75mm down from the underside of the existing footing. At least 24 hours after the base has been cast, a semi-dry 1:3 mix of cement and sharp sand is thoroughly rammed into the gap. A minimum of 12 hours should be left between the dry packing being inserted and an adjacent excavation being started

Figure 9.22: Typical sections through traditional mass concrete underpinning

a) Dry packing

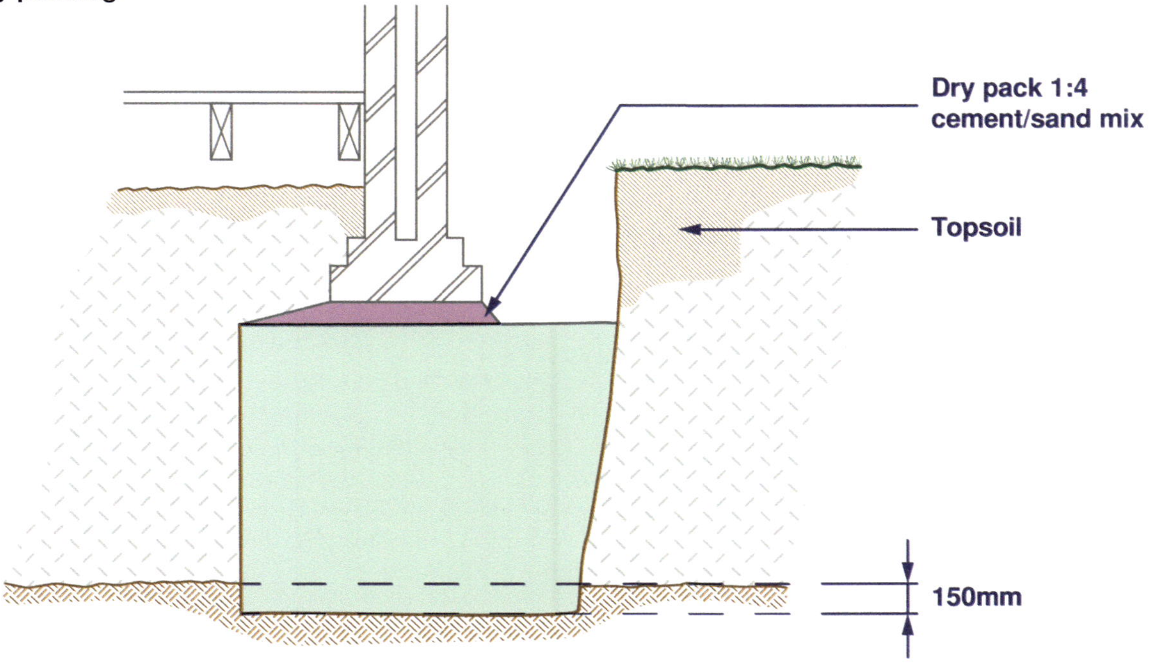

b) Concrete surcharging

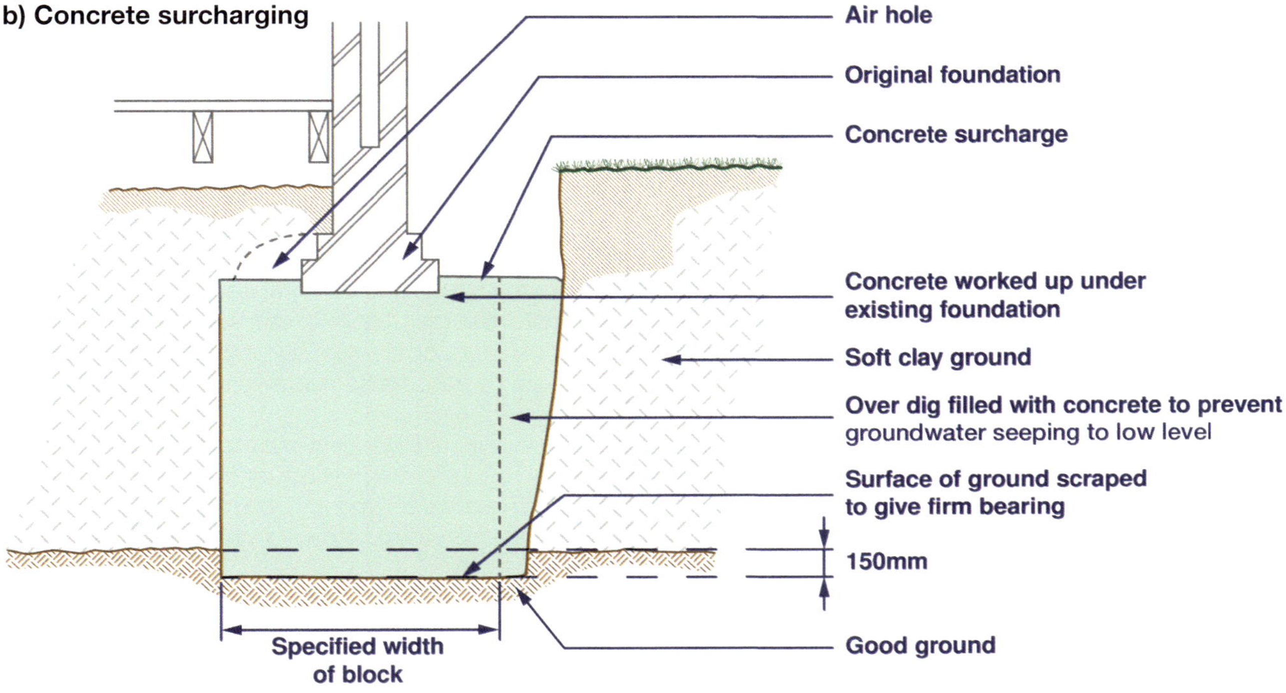

  - o Concrete surcharging (Figure 9.22b). This method is very seldom used, though described here for completeness. It is not a method that is recommended due to the possibility of air pockets and the shrinkage that occurs between the new and old foundation junction. The underpinning concrete is brought up to at least 150mm above the underside of the existing foundations. The concrete is 'worked' thoroughly under the old foundation by the vibrating poker to achieve a good percentage contact between the old foundation and the new underpinning. An air pocket vent left at the back of the old foundation assists in achieving good contact on the interface between the foundation and the underpinning. Again, a minimum of 24 hours should be left between casting and load imposition. The risks of not achieving adequate support to the existing foundation, brought about by trapped air must be guarded against. Great care must be exercised when preparing the underside of the existing foundation and in the forming of the voids for air pockets beyond the width of the existing foundation
- Concrete used is typically C25/30 with a 20mm aggregate and GEN3 offering sulphate resistance
- It is advisable that the whole of the excavated section be filled with mass concrete for its full plan area. Working spaces filled with stone can provide a route for groundwater to soften and undermine the underpinning, particularly in clay strata. Granular fill also provides an easy path to the base of the underpinning for future root penetration. Backfilling with excavated material is possible if care is taken to compact the material so as to prevent water penetration, but the procedure carries risks
- High strength grout; specialist high strength, free flowing, non-shrink grouts are available to fill the interface between the structure and the underpinning blocks
- As adjacent blocks are excavated, it should be possible to check the effectiveness of the pinning or contact surface between the new work and the original foundation
- Where partial underpinning is used, consideration should be given to providing transition blocks to reduce the sudden discontinuity of foundation depth at the end of the underpinning. This is intended to reduce damage from secondary movement
- Services below ground should where possible be accurately located prior to any ground works commencing
- The specified length of each block excavated should not normally be greater than 1m. No more than 1/3 to 1/2 of wall length should be unsupported at any one time. Localised concentration of loads should be identified and block locations and sizes adjusted accordingly
- The excavation for each block should be taken down to at least 150mm into a suitable bearing stratum, and a basic guide of 600mm below any live roots found. In stiff shrinkable clays, the use of slip membranes at the sides of the excavation are beneficial in preventing frictional adhesion between the concrete and the clay. Those specified should provide effective 'slip' capability and not be merely a token provision
- The adequacy of the stratum should be approved by the Engineer and agreed with the building control body
- Where the load bearing onto the underpinning is not uniform, e.g., where there are doors/window openings within approx. 600mm above the top of the underpinning, individual blocks should be keyed together using rebated notches, or another form of continuity provided in the sides of the adjacent blocks in order to help equalise differential loadings
- Individual pier or column bases are points of high load concentration and may require special precautions to be taken during any underpinning work. (Contractor also to bear this in mind during the actual works)
- All concrete should be well vibrated into place and it is recommended that a vibrating or poker with a minimum 50mm diameter be used. (Primarily a matter for the contractor)
- Continuity reinforcement or U-bars should be used where the full depth of a block cannot be cast in one operation (Primarily a matter for the contractor)
- Where encountered, all services should be examined for damage and replaced or repaired as necessary. Services should be sleeved (Figure 9.23) where they pass through the underpinning work to prevent damage from differential movement, but also protected during the casting to prevent damage by a large mass of fluid concrete poured against them. (Primarily a matter for the contractor)
- Any field drains encountered should be reconnected or rerouted as necessary. (Primarily a matter for the contractor)
- The underside of the existing foundation should be cleaned and any deleterious loose material removed. (Primarily a matter for the contractor)
- The foundation and any access holes or trenches should be adequately shored and strutted as work proceeds by the contractor, and edge protection offered by means of a guarded handrail system at ground level to prevent falls of debris and persons into the excavation
- When ready for concreting, the excavation should be inspected by the contractors' supervisor in every instance and the Local Authority/private building inspector. (Primarily a matter for the contractor but sample excavations should also be inspected by the Expert)

Figure 9.23: Drain service boxed out

- The sides of the excavations should be firm, vertical and parallel (Figures 9.24a and 9.24b). If there is a possibility that loose material will fall into the excavation and contaminate the concrete, polythene sheeting or other more permanent lining may be required. (Primarily a matter for the contractor)

Figure 9.24a: Underpinning base excavated

Figure 9.24b: Single pad foundation excavated and lined with Claymaster anti-heave protection

- For safety reasons, excavations should preferably not be left unconcreted overnight and never over weekends. Newly excavated ground left unsupported for more than a few hours can 'relax', dry out and collapse. It is also possible that the groundwater could leach into the excavations and soften unsupported sections of ground which may then be temporarily taking additional load. If, in exceptional circumstances, it is necessary to leave excavations unfilled overnight, they should be properly provided with planking and strutting to prevent collapse (Figure 9.25). Foundations should be temporarily propped. It is essential that excavations are protected such as to prevent accidents. (Primarily a matter for the contractor)

Figure 9.25: Example of inadequate shored and strutted foundation collapse

- Where the old foundation is temporarily propped overnight, the adequacy of the ground on which the propping is bearing must be carefully assessed, particularly as to whether it will prove adequate if the ground dries out or if water penetrates the excavation. (Primarily a matter for the contractor)
- Contractors may use both on-site mixed concretes and grouts, and ready-mixed concretes and grouts, depending on volumes, access, working space and economics. Irrespective of whether site-mixed or ready-mixed products are used, the standard checks to ensure compliance with specification must be carried out. These relate to mix designs, material specifications and risk assessments as well as quality control slump and strength testing

9.4.2 Guidance for Beam & Base underpinning

The following additional issues should be considered when the Expert specifies the extent of the works and the materials to be used for this type of underpinning:

- For this system, the structural design should be carried out by a suitably experienced Engineer, who may be the Expert or under the supervision of the Expert. If the design is carried out by the contractor, it should be reviewed by the Expert. A check by the Local Authority under the Building Regulations (Section 10.6 and Appendix E) is a worthwhile exercise but should not be relied on as a procedure for spreading risk
- Beam & Base underpinning is normally used where a firm stratum is found between 2–2.5m below ground level. Much depends on the Contractor and the excavation system employed as to the economic depth for this system
- Mass concrete bases are normally at 2.5–3m centres depending on the engineering design and structural arrangement of the building
- Excavation to be taken down at least 150mm into all adequate stratum approved by the Local Authority and the Expert
- Concrete mixes should be to BS 8500-1[78] and where used in the bases should be a minimum of GEN3, again with a recommended standard S2 slump class. The concrete specification for the beams will depend on the design but should generally not be less than C28/35 and preferably of higher quality. It is often better to have all the concrete of the higher quality, i.e., C28/35, to avoid confusion between different mixes on site. Durability and workability of the mix are as important as strength
- Depending on the precise details of the system used, the beam spanning between the bases will either sit on top of the bases or be accommodated in socket rebates formed in the bases at the time of casting. Temporary propping may be necessary where the beam is being installed in, or below, a particularly heavily-loaded wall. This is often in the form of stools or similar items placed at not greater than 1m centres along the line of the beam. These stools are normally sacrificial and are cast into the beam
- Beams can contain structural steel sections or be of a conventional reinforced concrete 'Bar & Stirrup' arrangement
- Adequate room must be left along the length of the beam so that the concrete can be properly placed and worked around the existing foundation or wall. (Contractor also to bear this in mind during the actual works)
- Inspection holes should be provided at the back of the existing wall to check that the concrete has adequately worked its way through to the back of the existing construction. (Contractor also to bear this in mind during the actual works)
- If concreting cannot be carried out immediately after excavation, the base of the excavation should be screeded over, or otherwise protected, so that the quality of the bearing stratum is not affected by the delay. (Primarily a matter for the contractor)

9.4.3 Guidance for piles with beams or rafts

The following additional issues should be considered when the Expert specifies the extent of the works and the materials to be used for this type of underpinning:

- All mini-pile and beam/raft arrangements should be designed by specialists in this field to the satisfaction of the Expert. If designed by the contractor, the design and detailing should also be to the satisfaction of the Expert. The Local Authority will require details to be submitted in either case
- Mini-piles used for underpinning are normally designed for an ultimate load of 5–50 tonnes depending on the pile diameter and the factor of safety specified. For a 100mm diameter pile with a factor of safety of 2.0, a working load of 5 tonnes is normally easily achievable in reasonable quality ground. A safety factor of at least 2.0 is advisable
- Mini-pile diameters can be up to 350mm, as referred to in the ASUC guidelines[79]. The size, type and depth of pile should be established in accordance with current standards, to establish structural and geotechnical capacity.

Driven piles have different installation constraints from augured or jacked piles. Driven piles are installed to a predetermined 'set'. It is important that some piles are re-driven in the presence of the Expert to confirm their adequacy before filling with concrete. This is particularly so if the piles have been left unfilled for some time after driving or where pore water pressure could affect driving resistance. Alternative methods of design verification such as pile testing are acceptable

- Augured piles are best used where vibration of the ground is likely to be unacceptable. Unlike driven piles, they are not 'load-tested' as they are installed to a predetermined depth and consequently their installation procedures should be more rigorously controlled and a more thorough site investigation may be required. When installing uncased bored piles alongside existing foundations, the risk of over-lifting should always be considered. This occurs when the rate of auger penetration is less than the pitch of the auger flights, and soil can then be drawn in from the side and up the flights, potentially undermining the existing foundations. This is a greater risk when attempting to penetrate dense bands of material using small, less powerful machines.
- Testing of piles is available using relatively simple methods. Dynamic/statnamic testing can be used to verify the load capacity of a pile, and sonic integrity testing will confirm whether a bored pile has any issues with discontinuity of the shaft (overbreak or necking) and will identify cracking and more serious anomalies
- Where an end bearing pile is to take support from the rock head, the pile should be socketed into the rock
- The verticality, line and position of mini-piles need to be constantly monitored. Checks should be carried out on randomly selected mini-piles — generally in accordance with the requirements of BS 8004[77]. Mini-piles which are found to be out of tolerance vertically, or bowed beyond permissible code limits should be carefully assessed as to their likely load carrying capability and resistance to buckling in the existing ground conditions. They should be tested and proved or, in cases of doubt, replaced.
- Needle beams through walls spanning between piles should not normally be spaced at greater than 2m centres. However, this can be increased if the wall is considered able to span or has been reinforced accordingly
- Where rafts are 'toothed' into existing walls, provision must be made for proper forming of the rebates, placing of the reinforcement or steel beams, and adequate compaction of the concrete to fill the rebate. Dry packing can be used above the rebates to achieve proper load transfer between the existing structure and the underpinning. (Contractor also to bear this in mind during the actual works)
- If, after or during driving, a bottom driven pile is found to contain water, this should be viewed with concern as it might indicate weld or plug failure which could affect pile set, and thus the pile's adequacy may be questionable
- If a pile contains water but is stable and dips the correct length, this water should be removed or concrete should be placed using tremie methods to displace the water without compromising the strength or integrity of the pile
- If dry packing is not used to achieve load transfer between the existing and new construction, then surcharging with fresh concrete can be just as effective, provided the operation is carefully executed and supervised.

9.4.4 Guidance for jacked piles

The following additional issues should be considered when the Expert specifies the extent of the works and the materials to be used for this type of underpinning:

- It is essential that accurate and thorough site investigation is carried out before any underpinning operation commences. It is important to predict the depth to which the piles will be jacked so that any anomalies encountered during their installation can be recognised. There are two principal types of pile used in jacking:
  - Precast concrete segments connected with mechanical joints. Sections can be up to 1m in length
  - Hollow steel tubes. Again, up to 1m long sections are normally used. Tube wall thickness is normally 3–13mm, depending on the jack load to be applied and the predicted ground resistance, and they will be typically jointed using either an internal or external sleeve joint. Care must be taken not to crush or 'concertina' the tubes, and it is essential that all faces are machined perfectly square
- The spacing of jacked piles depends on the load to be carried but they are normally set at approx. 1m centres. Consideration should be given to the piles having adequate lateral stability, to prevent the pile itself buckling and so that there is no possibility of the pile head moving sideways
- Static load tests are a simple matter with this type of pile as the jack and kentledge arrangement is in place to install the pile. Typically the pile will be tested by introducing a load cell between the jack pile rams and the pile; the load can then be applied and held for a predetermined period and any deflection measured. A full ICE static load test[80] can be undertaken by applying the test load in a specific cycle — but this is a test which requires two 12-hour cycles (typically over two days) so for subsidence repair piles a shorter period is often specified

- When the Expert and contractor are satisfied that a pile is approaching its design load, a stillage frame is placed between the pile head and the underside of the foundation to lock the load onto the pile. The jack can then be removed and the hole concreted. The stillage frame remains embedded in the final works
- Pile jacking procedures leave themselves open to many forms of abuse; it is important that supervision and inspection of pile jacking is undertaken frequently and thoroughly. (Primarily a matter for the contractor)
- Access pits should be excavated at the pile positions. It is important that the pile head is carefully held in place so that an even load is applied to the top of the pile. It is essential that the pile is jacked into the ground vertically. (Primarily a matter for the contractor)
- The base of the wall or slab against which the jacking takes place must be level to achieve an even application of load to the head of the pile, and sufficiently firm so that no local crushing or breaking of the original foundation material occurs. If necessary, a metal or concrete padstone should be fixed or cast immediately above the jack position to prevent over-stressing
- As jacking progresses, new pile segments are placed on top of preceding ones until the predetermined maximum jacking load is achieved without any movement of the final segment. For verification it is considered good practice to hold the maximum load on every fifth pile for a minimum of 10 hours (or overnight) and for 24 hours on every tenth pile. All jacks used should have a current calibration certificate

9.4.5 Contractor proposals, workmanship and quality control on site

Responsibility for achieving the specified quality of workmanship ultimately rests with the contractor unless the Conditions of Contract say otherwise. But, as reasonably appropriate and on a sample basis, the Expert should also be prepared to inspect the general standard of work during the course of the contract.

Notes from the previous sections apply as appropriate. Particular points to bear in mind include:

- With any underpinning, piling or polymer injection works it is essential to have experienced, trained and competent specialist contractors and tradespeople carrying out the work
- All contractors should price on the basis of the scheme as specified. If the contractor wishes to submit an alternative scheme, it should be presented separately and be accompanied by a full engineering specification and sufficient other details to enable the Expert to decide on the most appropriate scheme
- Where any alternative scheme is selected, full calculations, drawings, etc., should be provided, for approval by the Expert prior to commencing construction
- Agree the extent of the working area prior to the contract commencing and clearly define any 'out-of-bounds' areas
- Where premises are vacated by the occupants, all areas out-of-bounds to the contractor should be locked and preferably sealed against dust with masking tape. Carpets and underlay in areas to be trafficked should be taken up and stored off site
- Protect drives, walls, decoration, fixtures and fittings, lawns, etc. if liable to damage (Figure 9.26).

Figure 9.26: Example of plastic sheet protection over retained face brickwork and garden shed to prevent concrete damage

- Adequate welfare provision should be made by the contractor to comply with CDM 2015[15]. Due to the nature and duration of the works this should comprise either a standalone welfare unit, or an unoccupied area of the site that has been converted into a full welfare/site office area. Any areas should be kept in a clean state during the works and left thoroughly clean on completion
- Always hand dig in areas where services are anticipated. The use of tracking devices is essential
- Over-digging of holes for mass concrete underpinning can result in appreciable additional cost in terms of extra concrete. It is important to keep any over-digging to a minimum where possible. Filling of over-dig with granular material may not be permitted
- Dry packing between underpinning and the existing structure should be carried out thoroughly and by experienced tradespeople. A number of instances have occurred where underpinning work has failed because of the inadequate quality of the dry packing
- Welding of pile sections must be carried out by a proficient tradesperson experienced and qualified in welding
- Continuity reinforcement should be provided between piles and any beams or raft construction cast on the piles. Ideally, piles should have at least one reinforcing bar for their full length, particularly where piles may be required to take tension. Where full length reinforcement does not exist and tension in the pile is not anticipated, continuity reinforcement, usually a T12 bar, should overlap the steel casing by at least 1m and should bond into the beam/raft for the minimum bond length
- The material used in any void formers to separate underpinning work from surrounding ground or adjacent structure should be inert. Very low density polystyrene is probably the best material, but whatever material is used, it should not give off obnoxious or dangerous fumes during any subsequent decomposition. It must be capable of achieving the required amount of compression without the necessity for exceptional force. Any compressible layer should be properly jointed between adjacent sheets so there is no bridging of the void layer by solid material filling any gaps. The void former system should be carefully installed so that the anticipated movement has been comprehensively catered for. It must also be strong enough to support the construction loads placed on it
- Mini-piles will normally be 300mm diameter or less and fall into two categories:
  - Displacement mini-piles may be either driven, vibrated, jacked or displacement augered and do not bring arisings or spoil to the surface
  - Replacement mini-piles are gradually installed by boring, augering or drilling methods, and will produce spoil at the surface which will probably require appropriate off-site disposal
- Displacement mini-piles generally comprise permanent steel casings filled with *in situ* concrete with the required reinforcement. They may also comprise precast concrete, or steel, either tubular or I-section
- Bored mini-piles can be installed using tripod rigs, while augered and drilled piles will be installed using mini rotary hydraulically-powered rigs
- In ground which is heaving or suspected to heave, mini-piles are constructed using either teflon-coated, cardboard or plastic tubes on the outer surface of their upper levels to mitigate heave forces caused by swelling clays
- Mini-pile products and installation methods are constantly being updated and developed and care should be taken when dealing with new techniques and systems
- The Expert should require that a record of the locations and depths of any underpinning and piling installed be maintained, including any variations between the works actually carried out on site and those originally proposed. This record should be provided to the Expert by the contractor on completion of the works
- For designated or designed mixes to BS 8500 it is not necessary to take concrete test cubes, as this is the suppliers' responsibility. Consistance identity testing (slump test) should be carried out to ensure the flow or stiffness of the mix is as ordered, and appropriate for the intended use

9.4.6 Guidance for geopolymer injection

The following additional issues should be considered when the Expert specifies the extent of the works and the materials to be used for geopolymer injection:

- To assess the suitability of the system for the specific site, a ground investigation must be conducted by suitably competent and experienced persons, to assess both the geotechnical and geo-environmental status of the site and its vicinity. Ground investigations should follow the guidance of BS 5930[52], BS 10175[81], BS EN 1997-2[82] and its UK National Annex[83]
- To assess the suitability of the system for the specific site, a structural investigation must be carried out by a suitably competent and experienced person/s, to identify the nature and cause of any disturbance found, a description of the construction system and the nature of the construction materials — taking into account the intended purpose of the structure and any particular constraints relating to its operation. The vertical load on the

foundations must likewise be determined accurately, and also the stress state prevailing in the sub-surface. Soil treatment by injection of geopolymer carries a risk of inducing cracks in the concrete or masonry structure. As part of the structural investigation, the likelihood and consequences of such cracking must be ascertained and where necessary, mitigated against as part of the design process
- Treatment zones must be determined according to the soil characteristics and the configuration of problems identified in the geotechnical investigation and the structural assessments. The soil improvement targets (bearing capacity, deformation modulus, etc.,) must be determined on the basis of the results of the geotechnical survey and structural assessments (vertical load on foundations). Based on the results of the geotechnical study, structural assessments and the improvement targets, the injection point grid (number of injection levels, spacing between points, etc.,) must be determined by a specialist contractor
- Before starting any procedure, the location and nature of any buried pipework present in the treatment zones must be established by a competent Surveyor. The minimum distance between pipework and an injection point must not be less than 500mm
- While injecting close to drainage pipes (wastewater, sewage and rainwater), continuous monitoring by camera must be undertaken throughout. There may still be a risk of damage to drainage. Any damage must be detected, and the necessary repairs undertaken
- For sensitive pipework (e.g., gas pipes or optical fibre ducts) in the treatment area, the minimum distance of injection points must not be less than 2m unless the pipework concerned has been neutralised (sealed and purged) or diverted prior to initiation of the work. The pipework cannot be reinstated until it has been checked for correct operation by a specialist contractor and repaired where necessary
- Each injection is accompanied by laser level monitoring in order to detect any vertical movement of the treated structure. Precision is ±0.5mm. The equipment consists of a laser transmitter and several receivers, installed on the structure. The receivers detect slight altimetric variations in relation to the fixed horizontal plane traced by the laser beam of the transmitter. The transmitter is positioned a distance away from the injection site, at a neutral point not affected by the injection
- Any movement in the structure is measured in real time. All measurements are recorded in the site report. Any evidence of lift demonstrates that the treatment is effective. Expansion of the geopolymer in the first instance exerts pressure on the areas of least resistance, inducing horizontal compaction of the ground. Once equilibrium is obtained between the principal horizontal and vertical stresses, expansion of the resin continues, exerting vertical pressure (opposite to that caused by the overload and weight of the overlying ground or structure)
- The verification of ground improvement targets (i.e., specified bearing capacity etc.,) should be determined via the execution of post treatment standard *in-situ* geo-mechanical tests such as dynamic penetration testing (DPT), standard penetration testing (SPT) or cone penetration testing (CPT)

9.5 Partial underpinning

Partial underpinning is the provision of additional support to only part of the foundations of an existing property or interconnected group of properties effectively comprising one building e.g., a terrace of houses or a pair of semi-detached houses. There is nothing unsound with the principle of partial underpinning — indeed most underpinning is partial and is successful.

In the majority of cases, damage from subsidence, such as encroaching tree roots or leaking drains, may only affect part of a property. There is rarely any reason to extend remedial works beyond the affected area.

Some Engineers and Local Authority Building Control Officers have suggested that partial underpinning of a property may provide an uneven or unbalanced foundation. They argue that there may be a possibility that some differential settlement might occur in the future between the untreated and treated areas and cause damage.

In most cases foundations would not be expected to move in the future if they have not moved in the past, provided that:

- Any local problems have been rectified
- The structural arrangement of the property has not been altered
- The loading arrangement within the property is not altered
- The general ground conditions in the area do not change
- Adjacent trees are properly managed

The Owner has a reasonable entitlement to rectify their property so as to make it stable. However, in exercising this right, they might have to consider whether there are any obligations to neighbours. Clearly no actual physical damage should be caused to an adjoining property in the execution of any remedial works e.g., cracks in walls, broken windows, etc., nor must the work accelerate or initiate any subsidence that may be a threat to adjoining properties whether they be attached or not. The Party Wall, etc. Act 1996[14] provides for an adjoining Owner to agree to works proposed that are to a party wall or structure, or foundation works within a specified distance. In so doing, any proposal that seems to have such implications can be challenged before the works commence. The scope of this legislation therefore takes into account most of the eventualities that could occur with linked properties. Further details of this legislation can be found in Section 10.3.5 and Appendix C.

Care must be taken to ensure that the stability and serviceability of untreated adjacent properties are not compromised. In this respect, the mortgageability of the building, or of the different properties comprising the whole structure, may also have to be taken into account. It may be better to leave a group of properties to continue to subside in a uniform manner and so avoid major unserviceability of one or more of the group of properties occurring. The alternative being to stabilise one or more of the properties and accept the anticipated damage which may result from subsequent differential movement.

In instances where a building has been partially underpinned and sometime later subsidence occurs to another part which was not underpinned, Insurers normally deal with any resulting damage as a separate claim or, on occasion, as part of the original claim if there is continuing subsidence cover (Section 4.3.4). It is rare for existing Insurers to refuse to continue subsidence cover on buildings which have experienced subsidence if repairs have been properly designed and undertaken (Section 4.6).

9.6 Repairs to superstructure

9.6.1 General

Superstructure repairs may be all that is required because the Expert, after their investigations and surveys, has concluded that the cause of the damage has been mitigated by other means, such as tree management/removal or drainage repairs. In such cases, the property will neither have suffered any significant structural damage nor appear distorted to an extent which requires any rebuilding. In some cases, minor cosmetic repairs are all that is required (Section 9.6.6).

Cracking may have occurred due to a single occurrence e.g., where some adaptation or conversion work has been carried out on the superstructure which has caused the loads coming onto part of the foundation to be increased or affected to such an extent as to encourage slight initial settlement. Typical instances where minor cracking might occur but where no further continuation of major settlement results, include:

- Building an upper floor extension above existing construction, e.g., a new bedroom above a garage, or loft conversion
- Removal of a wall between adjoining rooms
- Enlargement of a doorway through a perimeter wall

Monitoring (typically over a 12 month period) will confirm whether movement is continuing and whether further action is required. Monitoring should confirm that the original decision not to do any ground stabilisation or underpinning works was correct.

If the property is unlikely to move any further, the question of repairing the superstructure arises. Structural movement does not always involve superstructure damage. If a building moves as a complete unit it may not suffer any superstructure damage. It might move downwards uniformly and so end up level, but at a lower level than it was originally constructed, or alternatively it may tilt.

Such uniform movement will normally only require repairs to drains and other incoming services unless the amount of tilt in the property is unacceptable. What is considered acceptable is subjective. Many older properties have quite exceptionally distorted or uneven floors, walls, etc., and such distortion is often considered to add character (and sometimes value) to historic structures. Tilt may add a quirkiness to more modern buildings (Figure 9.27).

Most buildings over 100 years old have some degree of distortion and it is rare that sloping floors or distorted walls cause concern to occupiers or Owners of these properties providing the properties are stable. From a practical point

Figure 9.27: Distorted brickwork panels (not subsidence damage)

of view, a slope in a floor of up to 50mm over a distance of 4m is probably quite acceptable, even if not in conformity with modern building standards or expectations of occupiers of more modern dwellings. Corresponding distortions in doors, walls, etc., are rarely complained of by those who own older buildings. When a room is furnished, only the most pronounced slopes tend to be noticeable or prove troublesome. Nevertheless some repairs to the superstructure may be essential to maintain the structural integrity of the building, and other repairs may also be necessary to rectify serious distortion that is unacceptable from a practical and aesthetic point of view.

9.6.2 Cracks in brickwork

These normally occur along or through the mortar joints, typically at the weaker points of a structure in or around door openings. They should at least be properly pointed, so as to maintain the weather proofing of the structure and prevent further deterioration. If considered necessary from a structural point of view, the brickwork may need to be bonded across and there are several methods to achieve this:

- **Repointing of cracking.** The most basic form of repair is to deeply rake out the crack line following the mortar joints and repoint in a similar strength and colour of mortar as the original. In the case of older properties and historic structures this should be undertaken in a lime mortar and if necessary a mortar analysis undertaken of the original to determine the strength of the original mortar as not to provide too hard or ridged jointing material. With more modern brickwork dyes, a good colour match should be achievable
- **Brick stitching.** Where there are multiple cracked bricks along the line of cracking, these can be cut out and similar new or reclaimed bricks stitched in to replace them; this also reinstates the bond and integrity back into the wall structure, as well as providing an aesthetic repair. This method of repair is also favoured in older historic properties as it maintains a similar brickwork strength along the wall and does not induce any localised stiffer sections

- **Resin bonding.** This method of repair is commonly used in more modern masonry, and following deep raking-out of the joint, a thixotropic epoxy resin is pumped into the open joint by means of a hand-held skeleton gun to bond the cracked masonry together. The resin is left back from the original mortar surface by around 20mm. Following full curing to prevent mortar discoloration, the joints are then repointed with a suitable colour-toned mortar to blend the repair into the existing
- **Bed joint reinforcement (often referred to as 'helical bars').** These stainless steel twisted bars, typically 6mm in diameter, but available in larger sizes if needed, are grouted into a raked-out horizontal mortar joint across a crack location to assist in tying the brickwork panel back together. The standard spacing of the bars is 450mm vertically and most bars are 1,000mm long, centred on the cracking. They can also be designed to provide masonry beams across large openings to increase stability and alleviate loading onto a lintel, or to tie back a section of an elevation. Smaller basic repairs can be undertaken by a competent builder whereas longer bar repairs, masonry beams or tying works must be undertaken by a specialist repair contractor who has been trained and has experience of the bar system

Where walls lean out of the vertical, their stability is rarely in question provided they are adequately tied back into the structure at roof and floor levels. Restraint straps, tie bars and externally-driven helical ties are commonly used for this purpose. Walls which are unacceptably out-of-plumb, e.g., 75mm in 3m, may need to be taken down and rebuilt, but again tying-in may be all that is required. Reconstruction should only be undertaken if it is absolutely necessary because a vertical wall forming part of a building which is otherwise obviously slightly sloping can be aesthetically unpleasant. This may be more unacceptable to an Owner or prospective purchaser than if the whole building or part of a building is slightly out-of-plumb.

Masonry walls, particularly those constructed using lime mortar, can accommodate appreciable distortions provided they are adequately tied into the structure. The criterion is not usually the extent to which the wall is out-of-plumb or bowed, but whether it is likely to deteriorate further after adequate tying or other remedial works.

Cracks identified as being due to wall tie failure should only be repaired once the defective wall ties have been replaced.

9.6.3 Floors

The relevelling of floors can be problematic. It is advised that floors should not be relevelled unless the degree of slope is genuinely unacceptable or there is another significant reason for re-levelling. Steps at doorways, conspicuous out-of-level between window sills, radiators, skirting-boards or similar fixed points of the structure can often be aesthetically unpleasant.

A suspended timber floor may be levelled either by laying additional material on top of the existing floor or by placing firrings on the existing joists, and reboarding. Undertaking the former can produce steps at doorways and staircases unless those positions are fortuitously the highest points in the room.

For solid floors, self-levelling screeds can be applied, or the original screed removed and replaced with a new screed run to the surrounding levels. Lowering a floor can cause major problems such as the taking down of the ceiling in the room beneath, or at ground level, the possibility of taking the floor below the DPC level. Groundbearing floor slabs can be re-levelled by the injection of expansive geopolymers beneath them on a grid pattern.

Rectification or replacement of staircases can be particularly troublesome and should be avoided if at all possible. Before undertaking the work, consider whether the final state will be worse than the original. Often a compromise is the best solution, where all or some of the slope remains, but for all practical purposes is acceptable.

9.6.4 Repair of plasterwork to walls, ceilings, etc.

These tend to be repairs of a more cosmetic nature, but if they are not done properly then seasonal cracking can re-occur, and indeed it may be sometimes unavoidable. Providing scrim across cracks in ceilings and metal lathing beneath new plasterwork over cracks in walls can substantially reduce the possibility of the recurrence of hairline seasonal cracking. No matter how skilled the tradesperson, it is impossible to achieve a perfect repair as shrinkage cracking is likely to occur. Embossed or heavily patterned wall and ceiling papers are often better than flat papers for covering minor flaws in repaired plasterwork. Common sense should always prevail in such circumstances.

9.6.5 Door/window realignment

Unless doors hang vertically on their hinges, they will not remain stationary when partially open (they will tend to swing open or closed). Sometimes doors and their frames distort as a unit but remain square. In other instances a door frame will distort out of square and the door no longer fits its opening. Small distortions of either sort are normally tolerable, but on occasions it is necessary to take off not only the door but also the architraves and door frame or lining and realign it correctly if it is to function efficiently.

Window realignment is rarely a sensible option, as reinstated windows are often conspicuously out of alignment with the brick coursing of the walls in which they are located. This should not be attempted unless absolutely necessary.

9.6.6 Cosmetic repairs

These would normally be carried out as part of conventional property maintenance, and include:

- Internal redecoration
- Filling to minor cracks to walls, ceilings and cornices, and around doors and windows
- Reinstatement of paths and landscaping

No further detail is provided on these types of repair, other than to emphasise that expectations should not be unreasonably high. The standards to be achieved should be agreed beforehand with the Owner, the Expert administering the contract and the contractor. It is sometimes useful to agree on a sample of workmanship before the full job is undertaken.

10 The contract and managing the project

10.1 Introduction

Previous chapters have identified and explored the various remedial works options available — ranging from arboricultural work, through to cause rectification, ground stabilisation measures, and superstructure repairs. This chapter provides information on the organisational aspects of managing a subsidence rectification project, including the placing of a construction contract and its administration, and the preparation of reports.

10.2 The role of the Expert

It is important that an experienced Expert is appointed. Their duties (Section 2.1.3) should include:

- A thorough investigation of the problem
- Design or approval of remedial works
- Administration of the contract

The Expert who designs the remedial works should consider the environmental impact of each rectification scheme, and the practicality and safety of the design during construction. The Expert who administers the contract should have a sound knowledge of general building construction, expertise in structural repair works and experience in this type of activity. This knowledge and expertise should extend to the types of contracts used, as well as the practical aspects of the repairs being undertaken.

Note, that asking the contractor to carry out unreasonably difficult tasks may incur additional costs and require additional inspection or supervision.

While it is not the Expert's role to supervise the contractor directly, they should exercise overall project management to ensure that the contractor is working safely, efficiently and is generally adhering to the design and specification, together with agreed timescales.

The contractor is responsible directly to the Owner for the standard of the work and should be competent and experienced in the field of work which is being undertaken.

As with all professional appointments, Experts should be engaged for their services on the basis of a written agreement set out in clear terms, together with the details of the responsibilities and roles of the Expert and the client, including arrangements for payment of fees. It is best if such an agreement is based on one of the standard forms published by such bodies as the Association for Consultancy and Engineering (ACE) or the Royal Institution of Chartered Surveyors (RICS). Ad hoc agreements not based on an accepted standard form can lead to misunderstandings.

10.3 Contract administration and procedures

10.3.1 General

A contract is an agreement between two or more parties which is usually legally enforceable. Subsidence remedial works contracts are in principle no different to any other building contract. The Employer (usually the Owner) requires building works to be carried out by the contractor and enters into a contract with them. The Insurer may provide a substantial amount of the money for the works but is not normally a party to the contract. The majority of subsidence claims involve work to domestic properties and this can present particular difficulties. Subsidence remedial work is invariably an imposition on the occupier; it has rarely been anticipated and subsidence is frequently discovered only when a house is being sold. It can involve the Owner in unexpected major financial outlay and in considerable unanticipated delay (or even loss) of the sale.

Most Owners rarely have anything approaching the scale of works of a subsidence repair contract carried out on their property. As a result, few know what to expect. Frequently, misunderstandings occur between the various parties due to lack of a procedure for dealing with the implementation of the contract. The situation can be further complicated by those not normally directly involved such as Solicitors, Local Authority Building Control Dept, Valuation Surveyors, etc. Chapters 2 and 3 detail the roles and concerns of the majority of these parties.

Where the proposed remedial works are extensive and very disruptive, it may be necessary for the occupiers of the property to move into temporary accommodation. This is frequently advisable for occupants who are elderly, disabled, have medical issues that are exacerbated by dust, and also where one or more essential services such as a kitchen or bathroom are rendered inoperable by the works. However, there is much to be gained on the point of security and general control of the contract if the occupier can remain in the property. Most underpinning and remedial works contracts are carried out with the occupier still in residence, unless the schemes are so intrusive that the complete ground floor needs to be disturbed.

The following subsections suggest what might be involved at various stages in the project.

10.3.2 Pre-contract stage

This is the initial stage. The Insurer has agreed with the Insured that a valid claim exists and remedial works are to be undertaken to restore the property to pre-loss condition. However, with the exception of the initial appraisal (Chapter 6), the site investigation (Chapter 7) and any mitigation measures, it is likely that nothing else has been done at this time.

Historically, the Insurer would appoint an Engineer/Loss adjuster that is suitably experienced in dealing with subsidence remedial works. However, more recently, and certainly for smaller less technically complex claims, a claim handling company assumes this role. Regardless of who has been appointed, for the purpose of this chapter, we shall deem them all as the 'Expert'.

The pre-contract site investigation, exploratory work and design of the remedial works is invariably undertaken by the Expert. The extent of the work is agreed with the Insured and the Insurer or their respective representatives. Note, that Insurers do not pay for preventative repairs (including the cost of tree management) except at their discretion. Depending on the extent of works, the site works may now take one of two approaches. The traditional approach and one that is still followed for larger more complex projects is that the Expert will prepare the tender enquiry documents for issue to contractors for pricing. These documents should particularly cover the points in Sections 9.1.3 and 9.6.4. For smaller, less technically complex claims, the project may be sent to a nominated repair contractor who has completed extensive prequalification vetting for technical competence, health and safety performance and has agreed to work for the Insurer's schedule of rates. This streamlines the process by removing the whole competitive tender aspect. While there are time-saving benefits to this approach, some critics are of the view that it removes the ability for a contractor to adjust prices to compensate for site-specific requirements, site locations and the adequate supervision and management of the site works.

It has become common for Lenders and prospective purchasers to expect that a guarantee or warranty is provided for basic structural repairs and decorations, and it is widely expected that it would be provided for extensive repairs and any form of underpinning works. If such assurances are required of the contractor or the Expert, their nature and scope should be agreed as early as possible, and any contractor requirements included within the tender documents. Prior to appointment, the contractor will be required to give full details of any guarantees which may be issued so these can be checked by the other parties. This also applies to the Certificate of Structural Adequacy (Section 10.8 and Appendix B).

10.3.3 Design administration stage

All relevant information as to the extent of the work to be carried out should be clearly defined and described. It is preferable for the detailed scheme of work to be agreed between the parties, e.g., Expert, Insured and Insurer, before documents are issued to contractors for pricing. Any private works to be paid for directly by the Insured should be specified separately at time of tender, and certainly no later than at the pre-contract meeting (Section 10.3.5). The Insured should have the policy excess confirmed and the method of collecting the excess agreed.

A reasonably accurate drawing of the general site and the relevant parts of the property to be repaired should be prepared and issued. It should contain all the known (and assumed) site information, including trial pit or borehole details, plus services runs. Use of CAT and Genny services location equipment can prove advantageous, along with obtaining service plans from national suppliers. Any assumed (i.e., provisional) information should be clearly noted as such and, where possible, this should be kept to a minimum in order to achieve accurate and realistic tender figures. The contractor should clarify in the tender the extent to which they have relied on any provisional information, in order that any subsequent claims under the contract can be determined.

It should be confirmed whether the tender is to be let as either:

- Detailed scheme with specification — Expert design — generally adopted as an industry standard
- Performance specification — with contractor design
- Based on the contractor's survey and scheduling

The first of these is the more usual for a contract funded by Insurers and/or for larger projects, whereas the latter is increasingly common for lower value remedial works. The tender period should be realistic and generally not less than four weeks. Lesser periods are not fair to the contractor and can lead to inflated tenders. When the tenders are received, the Expert will assess them and report to the Insurer and Owner with a recommendation as to which contractor should be appointed.

Where underpinning works are designed by the contractor, and a Certificate of Structural Adequacy is desired, the Expert must review the technical design produced by the contractor in order that they can subsequently issue the certificate. It should be clarified whether the Expert or the contractor is to obtain the Building Regulation approvals. Payment of Local Authority fees (by either the Insured or the Insurer) should be agreed as early as possible.

The Expert will generally be expected to design the engineering aspects of the works in accordance with relevant Codes and Standards. The Expert is not, except in unusual circumstances, responsible for the detailed supervision and organisation of the works. Those aspects are the responsibility of the contractor. The Expert should, however, make sure that the contractor is carrying out the work in accordance with the contract, and undertake all necessary inspections (Section 10.8).

The remedial works scheme designed by the Expert should avoid impractical and complicated details and requirements.

10.3.4 Contract and other insurance

Any requirements as to insurances to be expected of the contractor, Expert or Insured should be clarified at this stage, such as:

- JCT Minor Works Building Contract — Clauses 5.3, 5.4A, 5.4B and 5.4C, etc.[84]
- Professional Indemnity insurance requirements and public liability insurances
- Extensions of existing household policies, notification to contents Insurers
- Non-negligent damage cover resulting from the works

Specialist advice may be necessary if the inter-relationship between the various parties is thought to be complex.

If non-negligent cover is required, it is usually taken out by the contractor on behalf of the employer, i.e., the Insured, to safeguard against damage which might be caused to the employer's (Insured's) property which is not part of the contract works and to third party adjacent properties as a result of non-negligent action by the contractor during the works.

The contractor should have been made aware of this requirement via the tender documents so that an appropriate financial allowance can be made. Alternatively, it may be agreed that in the event of damage (or further damage) being caused through no fault of the contractor, the resulting costs should be borne by the employer, i.e., by the Insured, possibly under the umbrella of the overall claim. In practice this means that the Insurer undertakes to indemnify the Insured against liability for such damage as an inevitable part of the repair work.

It is normally necessary for the occupier separately to advise their contents Insurers that work is being carried out to the property because of the change of risk. They should also advise whether or not they will vacate the property while the work is undertaken, and specify what arrangements have been made for security, etc. Normally, contents Insurers will contribute to the cost of alternative accommodation.

Under the terms of the JCT Minor Works Building Contract[84], insurance cover under clauses of 5.4A or B should be put in place as appropriate. These relate to works and unfixed materials delivered to site in case they are lost or damaged. This is a standard form of cover with which most Insurers are familiar and can be affected by an extension of the household buildings policy for the period of the contract. Any extra cost will normally be borne by the building Insurer.

10.3.5 The pre-contract meeting

Although termed 'pre-contract', this is normally the first meeting between all parties after a contractor has been appointed, but it is usually held before work physically commences on site. Ideally this should be held at the property, with the Insured, Expert, contractor and, if possible and appropriate, the Loss Adjuster. Where works have been secured under a single source tendering process, this pre-contract meeting often takes place between the Insured and the contractor only.

At this meeting the Expert should explain in detail to the Insured, in the presence of the contractor, the extent and scope of the work to be carried out by the contractor and also which aspects of the contract are to be dealt with by sub-contractors.

It is not unusual for an occupier to decide that, as work is already being undertaken, it would be cost effective to arrange additional work at the same time (e.g., improved standard of decorations, the installation of new windows, or even the building of an extension). Such additional work must be clearly identified and programmed into, or running parallel to, the preparation of the remedial works contract, and ideally should have been included at pre-tender stage.

The occupier may well have had copies of drawings and specifications at an earlier stage, but it is often only by this 'hands on' approach involving the occupier that they begin to understand the extent of the work to be carried out. Delays and interruptions to the insurance-related works can involve cost to the contractor. The Insurer may not have any obligations to meet such additional costs, so they will invariably fall as a direct extra charge to be met by the Owner. A programme should be agreed. This is particularly important if alternative accommodation arrangements need to be made as private work will add to the duration of the contract.

At this stage the Experts should explain their role to the Owner. Their role in this respect should be acknowledged by all parties from this juncture if the contract is to run successfully.

It is also important to decide what temporary protection is to be provided to paths, doorjambs, etc. in order to reduce the necessity for reasonably avoidable remedial work. Facilities for the use of the contractor are to be agreed, such as power supply, water supply, toilets etc.

Where the Insured are to remain in residence for the duration of the works they should be made fully aware of the inevitable and unavoidable disruption, dirt and discomfort. Care should be exercised by all parties to avoid extending the limits of disruption beyond the area of the works.

The Expert should also explain that after underpinning has been completed, minor initial settlement might occur but that this is no cause for concern.

The Expert should explain the contractual relationship between the parties and clarify that the Owner is employing the contractor. Following this meeting a set of minutes should be produced and circulated, for agreement of all parties, by the Expert. The Expert's duty is to administer the contract impartially.

Party Wall Act (Appendix C)

The Party Wall, etc. Act 1996[14] requires the Owner to inform their neighbour when it is intended to undertake buildings works to either a party wall or to excavate close to an adjacent property. The intention of the Act is to protect both parties from spurious claims for damage made by one on the other.

The Expert should be fully aware of the details of the Act and apply the Act fully as appropriate. The duty of the party wall Surveyor is to the structure and not to either of the adjoining Owners, although the Surveyor will be appointed by either party.

Condition survey/photographic records
Irrespective of whether the Party Wall Act is relevant to a particular project, a condition survey with photographs should be carried out to assist in reinstatement and to capture the pre-work condition of the property. It is essential that sufficient photographs are taken of the whole site along with relevant descriptions. The condition survey must then be issued and agreed by all parties. The photographic condition imagery can then be added as rooms are cleared of personal items and additional damage becomes exposed.

Terms of payment
Before the works commence, a direct payment mandate is signed by the Insured. This enables payments once certified by the Expert to be made directly from the Insurer to the contractor. This gives the contractor confidence that fair payment for properly completed work will be forthcoming and that there will be no abnormal delays. Payments should be made to the contractor within the period stated in the contract. This arrangement invariably improves the smooth running of the contract for all parties.

As well as the contractor, the Expert is entitled to fair payment for their professional services. The Expert's fee may also be subject to a mandated arrangement with the Insurers. Payment procedures should also be agreed before the work commences or, if possible, even before the design work begins.

Interim certificates and payments should be processed at intervals stipulated in the contract (usually monthly) to enable the contractor to maintain a reasonable cash flow. Payment periods should be stated and not exceed four weeks. The Expert should clarify this point with the contractor and obtain their agreement. In practice, these certificates are normally sent directly to the Insurer or Loss Adjuster by the Expert, and the existence of a mandate facilitates this procedure. The Owner/Insured should be provided with a copy of all certificates.

On completion of the contract, the final account should be processed promptly and the contractor should expect to be paid promptly following certification by the Expert. Should there be any dispute, it is at this point (and not before) that the appropriate amount of disputed money be withheld while the matters in dispute are resolved. It is not fair or acceptable to the contractor that the whole amount of the final certified account be withheld, and this would usually be a breach of contract.

It will be agreed at the meeting how the policy excess is to be paid by the Insured. The most common method is for the contractor to raise an invoice to the Insured and collect the excess before work begins on site. This is then deducted off the final invoice certified by the Expert on completion.

In some instances, where the excess cannot be paid, and agreement to complete the works 'less the excess value' maybe reached e.g., not decorating a bedroom.

Any provisional sums included in the contract to cover possible unforeseen or undetermined works must be clarified and properly established at this stage so that all parties are aware of the implications to both the contract price and programme. Delays in clarifying provisional items can be disruptive to the contract and costly to all involved.

10.3.6 Contract stage

The signing of the contract should occur prior to commencement of any work on site. The contract for the remedial work is a binding and enforceable agreement between the Owner (the Employer) and the contractor and should in its simplest form be an agreement on what is being undertaken, when and for what cost. The Insurer is not a party to the contract but usually provides the funding for part or all of the agreed cost of the work being undertaken.

It is not uncommon for the contractor to ask for a payment mandate to be signed. This is a simple document instructing the Insurer to pay the contractor directly rather than via the Owner. This has advantages to both parties; the contractor has the comfort in knowing that a potentially large sum of money is not passing through the hands of the Owner, and if there is a delay in payment the Owner will not be pursued by the contractor directly.

A contract can take many forms, from a verbal agreement to a bespoke formal document. However, it is always preferable to have a written contract in place which is a clear record of what has been agreed between the parties and can be called on should there be a dispute or misunderstanding in the future. The JCT form of contract is the industry standard for building works in the UK. The JCT produce several variants depending on project size. However, for domestic projects the Minor Works Building Contract[84] is the most common. This formal contract even if not itself signed, may be referred to in a letter of agreement between the parties and becomes the rules under which the work is undertaken. Once a contract has been formally accepted it becomes a legally binding agreement which, depending on the scale of the works, may be a daunting prospect for the Owner but they should be able to rely on the advice of the Expert as to the appropriate contract to be used and the terms to be included within it. The Insured may, of course, obtain alternative advice if they are prepared to meet the additional cost of such advice.

Insurers may have a preferred contractor with whom they have negotiated rates (possibly based on them undertaking a large volume of work) and Insurers may ask the Owner to give them authority to appoint a contractor on their behalf.

10.3.7 Project management of the contract on site

Traditionally, the contractor would be responsible for serving notices and permissions prior to the start of the works. These could include matters relating to the party wall, statutory services or Building Regulations. They would programme the work and coordinate subcontractors and trades. Where Insurers have appointed more than one contractor directly, it is not uncommon for them to appoint an Expert to undertake the role of contract administrator to liaise with contractors, the Owner and third parties. This person may also be the Loss Adjuster, Engineer or contracts manager employed directly, but it is essential before embarking on a project to clearly establish the roles and responsibilities of all parties.

Good communication is essential particularly with the Owner, whether they remain in the property or vacate it. If they vacate, then an exchange of contact details is essential. However, once the contractor takes possession of the property the Owner should be reminded that it is a construction site and may be asked not return without being supervised.

A typical subsidence repair project has three phases; preparatory works, structural works, and reinstatement works, and will involve handing over between contractors and trades.

Cleanliness and protection of the property, including the external landscaping, paths, driveways, etc. is essential. Failure by the contractor to protect such areas will require reinstatement of any items which have been damaged or contaminated as a result of the works.

All parties must be aware that subsidence remedial work is essentially repair rather than new works, and therefore different expectations will apply. As with any repair, 'blending', 'matching' and 'marrying-in' of repaired and reinstated elements with original untouched elements has to be carried out carefully but the repair may still be evident. This is a concept that needs to be carefully explained to the Owner, as their expectations and what Insurers consider to be the limit of betterment are often mismatched.

The work involved in the repair usually takes two forms:

- Work in the ground. Frequently the Owner is not concerned personally about this work and the Expert normally liaises the contractor on the Owner's behalf
- Work above ground and cosmetic repairs. Invariably it is at this stage that the Owner takes greatest interest. It may be at this stage that some friction can arise as to the expectations of what is required if this has not been adequately explained and agreed at the pre-contract stage and/or the pre-contract meeting. The Expert should administer the contract fairly and firmly between the parties

If, during the course of the contact, the Expert is not satisfied with any aspect of the work being carried out by the contractor, the contractor should immediately be informed in writing and asked for a prompt reply as to the action proposed to rectify the matter.

The contractor is entitled to receive prompt payment of certified invoices. In turn Experts, Loss Adjusters and Insurers have an obligation to process them promptly. Contractors frequently complain that interim payments are unreasonably delayed because of the casual attitude of one or more of these parties. The contractor also relies on prompt instructions from the Expert.

The Owner should be encouraged to read and understand the contract they sign. The Expert should explain all the salient points. It is outside the scope of this guidance to discuss the terms of the contract in detail. Reference should be made to the appropriate forms of contract.

10.3.8 Work in the ground

The Expert usually deals with the queries as they arise during the day to day running of the project. During this stage of the work the Expert must be prepared to visit the site at relatively short notice and, if necessary, to make decisions promptly while on site.

Correspondence between the Expert and contractor recording variations to the contract should be made promptly, in accordance with the contract and conditions and preferably within seven days.

It is sometimes the case that the ground conditions encountered are not the same everywhere as those identified in the site investigation report. Even when a competent report has been prepared, it is, of its very nature, a record of examinations of the ground at several individual sample locations.

Ground conditions may be worse elsewhere. The groundwater conditions may alter if the works are being undertaken several months after the site investigation. This may cause variation to the contract works for which additional payments might reasonably be requested by the contractor. When delays occur due to problems in the ground, notification in writing should be sent promptly to the Expert by the contractor advising of any difficulties, why they have occurred and any possible contractual implications.

Where large cost implications arise due to unforeseen or unspecified work being required, an early resolution of such matters is essential, and the Insurers or Expert should be informed in a timely manner. From a practical point of view delays will invariably occur unless the Expert has been delegated authority to agree such matters. Equally, the Insurer must react promptly to problems of this nature properly notified to them by the Expert. It is essential to establish the extent and cost of any extra work (and who is to pay for it) as soon as possible.

10.3.9 Work above the ground

This is the stage of the contract where problems and frustrations can arise for all parties — often as a result of misinterpretation of what is required/expected of the contractor.

Firm administration of the contract by the Expert is essential to the contractor. The Expert should, when requested, promptly provide the contractor with clear directions as to what is required under the contract.

If not previously explained to the Insured, a meeting should be held to agree the extent and standard of the 'cosmetic works' through to completion. Insurers or their representative should attend this meeting, as to all intents and purposes it is they who approve the expenditure on such cosmetic works. On completion of the structural works and the attendant cosmetic repairs/reinstatement, and prior to decoration, the contractor is expected to 'sweep out' the property, clean off the paths and generally tidy up ready for redecoration. Note, that this 'tidying' is not expected to be of a standard commensurate to professional domestic cleaning. On completion of the total works, contract cleaners are sometimes employed to finish the cleaning, but this is usually carried out by the occupier.

10.3.10 Decoration

It is often necessary to carry out decorative repairs to a building that has suffered from subsidence and/or heave, whether it has been underpinned or stabilised. Sometimes it is necessary to carry out a complete redecoration contract, but this is rare. Insurers may look on excessive redecoration as 'betterment' and not provide funds to cover this — expecting it to be paid for by the Insured. That said, the likely extent of decorative repairs is often difficult to assess at pre-tender stage. Usually therefore, a provisional sum is included in the estimate for the works to cover the decorative repairs deemed necessary following stabilisation or underpinning works. This sum should be realistic and fairly represent the full extent of the work likely to be required.

The contractor that is executing either the underpinning or superstructure repairs, or both, has an obligation to see that existing decor, fittings and fixtures are adequately protected. A small degree of further deterioration of existing decorations during the works is tolerable, but negligent damage caused by the contractor is not. If this occurs, the contractor should be expected to contribute appropriately to reasonable redecoration costs.

At an appropriate point during the project, (usually on completion of the underpinning stage but before full-scale superstructure repairs commence) an assessment of the decorative repairs required should be made and agreed, particularly with the Insured.

The extent and standard of decorative repairs are often particularly emotive issues and an important part of the project management of any subsidence claim is to make all parties aware, before the work commences, just how the finished article will look. If, at this stage, the Insured decides to have improvement works carried out to the decor during the contract works, e.g., upgraded papers, paints and other finishes, then Insurers will normally require a financial contribution to be made by the occupier or Owner, equal to the difference in cost between normal reasonable rectification to the pre-subsidence damage standards, and the Insured's required improved standards. In other words, the Insured pays the difference between the two.

It is sometimes the case that the decoration is taken out of the main contract and the Insured is allowed to deal personally with this aspect of the remedial works. In exceptional circumstances it may be considered as an item allocated to offset part or all the excess contribution required to be paid by the Insured.

Alternatively, the contractor could be asked to carry out the decoration work in addition to the main contract, at which stage a firm price should be submitted once the extent of the work is known. It is usual for a separate decorating contractor to undertake this work, whether employed by the Owner or the contractor.

As with all other contractors, the decorator expects and is entitled to be paid promptly. It is also worth noting that smaller decorating firms often expect to be paid immediately on completion of the job. This is another reason for allowing the Owner to look after this part of the works which, if necessary, can be carried out at a later date.

10.3.11 Final completion

Towards the end of the project, a meeting should be held to discuss the final stages of the work. As stated in Section 10.3.5, those who attend or arrange this meeting differ slightly depending on who the client or Claims Handler are for the project. In a traditional tendered project, the Expert (whether that be an Engineer, Loss Adjustor or Contract Administrator) will attend the meeting with the client and the contractor. In the more modern single source or term contract approaches, the meeting will normally just be held between the contractor and the client. The purpose of the meeting remains the same; have the works been undertaken following the agreed schedule of repair? Is the client happy with the works undertaken and do any issues (snags) remain with the works that need to be resolved prior to completion? Where a dispute or major failing occurs, the Expert will act as a professional intermediary and offer their professional opinion.

It is not uncommon for the client to be asked to sign a 'satisfaction note'. By signing this, the Owner confirms their acceptance of the works to the Insurers, but the wording of the form should not reduce or remove the legal rights of the Owner to pursue any reasonable additional aspect of the claim which may subsequently come to light, nor should it affect their rights in law as regards any of the other parties involved. It should only be an acceptance in principle that the works appear complete to the Insured. The Insured and the Expert should check the wording carefully.

Following the completion meeting and or the receipt of the satisfaction note, the contractor will submit their final invoice — for processing by the Expert and payment by the Insurer. If during the liability period of the contract, any works come to light which have not been properly completed by the contractor, they must be attended to during this period. A retention payment may be held, in the event that the contractor is unable or unwilling to return. If any guarantees, warranties, or Certificates of Structural Adequacy are to be issued, these become due immediately on full payment, with each party being separately responsible for issuing their own documents in the agreed terms once they have been paid.

10.4 Health and safety

There are a number of regulations governing the way in which health and safety matters apply to building contracts. All parties have some degree of responsibility for compliance with the appropriate regulations.

For activities associated with subsidence rectification it may be necessary to comply with various regulations including:

- The Health and Safety at Work, etc. Act 1974[85]
- The Provision and Use of Work Equipment Regulations 1998[86]
- The Lifting Operations and Lifting Equipment Regulations 1998[87]
- The Management of Health and Safety at Work Regulations 1999[88]
- The Reporting of Injuries, Diseases and Dangerous Occurrences Regulations 2013[89]
- The Control of Substances Hazardous to Health Regulations 2002[90]

Of particular note are the Construction (Design and Management) Regulations 2015[15]. Where subsidence works are being carried out on behalf of a company or corporate body, e.g., an Insurer, or it involves more than 30 days' work with 20 workers at the same time, or will involve more than 500 person hours, the CDM Regulations apply. Furthermore, all design undertaken for the purpose of construction, falls within the scope of the CDM Regulations.

With the exception of duties always applicable to designers, because of the limited number of people involved and short duration of the works, it is unlikely that the project will be notifiable to the Health and Safety Executive (HSE) for the majority of subsidence remedial works, unless the project also involves work on a substantial part of the building. Where CDM Regulations apply, they require that the HSE be notified of the works in advance.

The CDM Regulations require that certain key appointments are made, and details of the responsibilities of these appointees should be understood.

Companies may, depending on their position in relation to the work, have one or more of the statutorily defined roles in the following list:

- Client
- Principal Designer
- Designers
- Principal Contractor
- Contractors
- Workers

The responsibility for pre-construction planning lies with the Principal Designer and this is passed to the Principal Contractor for the construction phase. These two posts are required to be by appointment of the client. The designated duties of each of the six roles are clearly defined and this information, along with guidance, is available from the HSE. Information on all other health and safety-related regulations including information on occupational health is also available from the HSE.

These legal requirements should be understood by all parties and should be incorporated within the contract.

10.5 The Housing Grants, Construction and Regeneration Act 1996 (also known as the Construction Act 1996)

Consideration may need to be given to the requirements of the Construction Act 1996[21].

The Construction Act does not apply to contracts between a contractor and a residential Owner/occupier. Thus, the contractor on an underpinning project for a domestic residential Owner/occupier is not bound by the Act. However, all other construction-related contracts are bound by the Act. Therefore, although the main contractor may be exempt in this situation, any contract they may have with a sub-contractor will be bound by the provisions of the Act.

All other construction operations including underpinning and superstructure repairs for non-residential domestic occupiers are applicable to the Act. The main provisions of the Act relate to prompt payment and the right to refer a dispute arising under the contract for adjudication.

10.6 Building Regulations

10.6.1 Scope of the regulations

Building Regulations are made under devolved powers to the Northern Ireland Executive, Scottish Government and Welsh Government; with the UK Government retaining responsibility for England.

Building Regulations applicable across the UK are summarised briefly in this section, with more detail provided in Appendix E.

10.6.2 English and Welsh regulations

The scope and the basis of the Building Regulations in England and Wales are discussed in Appendix E1.

Since 2011, separate regulations have been made by the Welsh Government; however, the regulations are similar to those in England and are considered together here.

10.6.2.1 How to demonstrate compliance with the regulations

In England and Wales, a choice of either a Local Authority or Approved Inspector can be made for the oversight of works covered by the regulations.

An Approved Inspector is an individual or organisation (often a private company) authorised to carry out building control work. They must be registered with the Construction Industry Council Approved Inspectors Register (CICAIR)[91] which maintains a list of registrants and oversees their operation through audit and monitoring. Note, that from April 2024, the term 'Approved Inspector' is largely superseded by 'Registered Building Control Approver' (RBCA)'; therefore any reference to 'Approved Inspector' should be read as a reference to a Registered Building Control Approver.

When using a Local Authority, submissions may be made through the building control approval with full plans or 'building notice' route. For both, regular inspections following a site service plan are undertaken to review progress of works and establish whether compliance can be demonstrated; ultimately leading to a 'completion certificate' if satisfactory.

For building control approval with full plans, the authority will undertake a review of submitted details and issue an approval (with possible conditions) if they are satisfied the details show compliance. Confidence in demonstrating compliance can be achieved if the works are in accordance with these approved details.

Limited details are given with a building notice, and no formal review or associated approval is given. Therefore, works could be undertaken that are not compliant, resulting in costly abortive or remedial works. Note, that the Local Authority can request any plans or details necessary for the works being undertaken under this route.

Where an Approved Inspector is being used, the person carrying out the work and the Approved Inspector will jointly give the Local Authority a notice advising that the works will be overseen by the Approved Inspector; this is known as an 'initial notice'.

The Approved Inspector will undertake inspections throughout the works and issue a final certificate on completion if they are satisfied that compliance is demonstrated. An Approved Inspector has no enforcement powers; therefore, if they are unable to resolve non-compliance they must transfer the application to the Local Authority for enforcement action, this is known as 'reversion'.

10.6.2.2 Building Regulations as they apply to foundations and underpinning

The requirements for foundations and other structural elements are contained in the following extracts from Part A of Schedule 1 to The Building Regulations 2010[16]:

Loading

A1. (1) The building shall be constructed so that the combined dead, imposed and wind loads are sustained and transmitted by it to the ground:
- (a) safely; and
- (b) without causing such deflection or deformation of any part of the building, or such movement of the ground as will impair the stability of another building.

(2) In assessing whether a building complies with sub-paragraph (1) regard shall be had to the imposed and wind loads to which it is likely to be subjected in the ordinary course of its use for the purpose for which it is intended.

Ground movement

A2. The building shall be constructed so that the ground movement caused by:
- (a) swelling, shrinkage or freezing of sub soil, or
- (b) landslip or subsidence (other than subsidence arising from shrinkage), in as far as the risk can be reasonably foreseen, will not impair the stability of any part of the building.

Regulation 3(1)(f) defines 'work involving the underpinning of a building' as 'building work' to which the regulations apply.

Regulation 4(3)(a–c) requires that work shall be carried out so that after completion, the building complies with the relevant requirements of Schedule 1; or, where it did not comply with any such requirement, is no more unsatisfactory in relation to that requirement than before the work was carried out. This is more commonly known as being 'no worse'.

Requirements C1 and C2 relate to resisting contaminants, the passage of moisture into the building (ground or otherwise) and damage to the fabric of the building by contaminants or moisture.

Requirements H1–H3 are concerned with drains and soakaways; they should be laid to protect the foundations.

Requirement H4 relates to the protection of buildings over or close to a sewer, as well the sewers themselves. Account may need to be taken of any sewers in close proximity to the works.

The original compliance with these requirements could be adversely affected by underpinning or remedial works to the foundations and therefore be a contravention of Regulation 4.

10.6.3 Scottish Building Regulations

The scope and the basis of the Building Regulations applicable to Scotland are dealt with in Appendix E2.

How to ensure compliance with the regulations:

- By compliance with the relevant standards set out in the supporting Technical Handbooks[92, 93]
- By conforming with provisions which are stated in the Technical Handbooks to be deemed to satisfy the relevant standards, or
- By any other means which can be shown to satisfy the relevant standards

Designs, materials and methods of construction must therefore be at least to the standards set out in the Technical Handbooks to satisfy the regulations.

The Handbooks give information on the typical application of the regulations and provide:

- A full list of defined terms used in both the Statutory Instruments and Technical Standards
- Instructions for the interpretation of references to other published standards
- A complete extract of regulations covering:
  - o Descriptions of building and work, including the provision of services, fittings and equipment not requiring a warrant
  - o Limited Life Buildings
  - o Measurements
  - o Fitness and durability of materials and workmanship
  - o Building Standards — construction
  - o Building Standards — demolition
  - o Building Standards — service, fitting and equipment
  - o Building Standards — conversion
  - o Provision of protective works
  - o Clearing of footpaths
  - o Securing of unoccupied and partly completed buildings
  - o Relaxations
  - o Building Regulations as they apply to foundations and underpinning

Structural requirements are listed in Section 1 of the Handbooks. The relevant criteria are:

- Loadings on the building
- Nature of the ground
- Collapse or deformations
- Stability of the building and other buildings
- Climatic conditions
- Materials
- Structural analysis
- Details of construction

This notes that loads used in calculations must allow for possible dynamic, concentrated and peak load effects that may occur.

There is no specific requirement relating to underpinning.

10.6.4 Building Regulations (Northern Ireland)

The scope and basis of the Building Regulations in Northern Ireland are discussed in Appendix E3.

How to ensure compliance with the regulations:

The Building Regulations (Northern Ireland) Order 1979[94] enables the Northern Ireland Department of Finance to make building regulations. The current regulations are The Building Regulations (Northern Ireland) 2012[18].

Building Regulations set mandatory requirements regarding the design and construction of buildings. 'Functional regulations' set requirements but do not prescribe a particular standard that should be achieved; the method chosen to demonstrate compliance is open for choice to the applicant or their agent.

The department publishes Technical Booklets[95] in support of the functional regulations that show methods and standards of construction which, if followed, are likely to demonstrate compliance with the regulations. Note, that these documents are not regulations in themselves and therefore do not set mandatory requirements.

The functional regulations are grouped into lettered subject parts, and the technical booklets in support of each of these parts follows the same referencing format. For example, structure is covered by Functional Regulations 30 & 31, which is Part D of the Building Regulations and guidance in support of these regulations is contained within Technical Booklet D10.20.

For assessment purposes, district councils are likely to benchmark alternative or innovative proposals against the guidance in the Technical Booklets.

'Building work' in the context of foundations and underpinning means 'the erection of a building, the structural alteration or extension of a building (including work in connection with a material change of use) or the provision of a service or fitting'.

There is no specific definition for work relating to the underpinning of a building within the Northern Ireland regulations; such work will tend to be considered under Regulation 7 in Part A 'Application to alterations and extensions'.

Building work shall be carried out so that it complies with the relevant requirements of the regulations. The requirements for foundations and other structural elements of a building are contained in Functional Regulation 30 (Part D) and guidance in support of achieving the Functional Regulation is contained within Technical Booklet D[96].

10.7 Guarantees

The majority of specifiers of underpinning ask for guarantees to be provided by the contractor. During the 1980s and early 1990s this was largely driven by Lenders via their Valuation Surveyors. More recently, this appears to have become the requirement of conveyancing Solicitors on behalf of prospective purchasers.

Guarantees provide perceived comfort and a feeling of security. However, unless insurance backed, these guarantees are only as good as the quality and stature of the company issuing them. They will rely solely on the company continuing to trade profitably during the term of the guarantee.

Guarantees have been provided for periods of up to 30 years. However, latterly, guarantees of 10 years have become more common, largely mirroring the guarantee period provided by major housing structural warranties such as NHBC and LABC Warranty. In practical terms 10 years has been acknowledged as a suitable period, as it is likely that if a problem occurs it will manifest itself within the 10-year period and usually within 3–5 years.

Note, that the Association of Specialist Underpinning Contractors (ASUC) has an Insurance Backed Latent Defects Scheme in place. A typical guarantee is shown in Appendix F.

However, in addition to the issue of the financial substance behind the guarantee, the terms need to be considered carefully. They will rarely reflect the contractor's legal and contractual obligations but usually qualify and limit their liability. In consumer contracts, certain legal rights cannot validly be limited or excluded, but in other cases the terms of these guarantees must be scrutinised to ensure they are expressly 'without prejudice' to other rights. If they are not, and they emanate from the contractor, they should probably be rejected.

In the few known cases when such contractors have had to reinstate allegedly defective work, it appears to have been adequately attended to by the particular contractor involved.

10.8 Certificate of Structural Adequacy

The following procedure is proposed and should result in a more meaningful assurance being given in the form of a Certificate of Structural Adequacy (Appendix B) from the Expert, particularly when an insurance backed guarantee is unavailable:

- The Expert chosen to carry out the investigation should decide what remedial works, if any, are required. If no major works are necessary, a Certificate of Structural Adequacy can be issued by the Expert at this stage
- If major remedial works are required, the Expert and contractor selected to undertake the work should be experienced and competent in the services and type of work involved
- The Expert should design the remedial works within the extent of the information available from the investigations
- The contractors invited to tender, including the one eventually selected to undertake the work, should be chosen because of their known competence or experience of the type of work required
- The Expert should carry out site inspections at sufficiently representative intervals throughout the contract in order to assure that the contractor is generally complying with the specification
- The Expert should require the contractor to rectify any unsatisfactory work detected, prior to issue of the certificate

On completion of the works, it should be possible for the Expert to issue the Owner with a Certificate of Structural Adequacy for the area of the property subjected to remedial works. Such a certificate would not be a warranty or guarantee, but it would be a considered professional opinion by an Expert. It would not apply to the cosmetic work of re-pointing, re-plastering, decoration, etc., but only to the major structural work which contributed directly to the stability of the building, such as the underpinning itself, and any tying into the structure of main walls, etc.

The certificate and any accompanying drawings, etc. will state:

- The name of the Owner and address of the property
- The damage which the property sustained in general terms
- The cause of the damage as established by investigation
- The property stabilisation measures (if required)
- The remedial works undertaken to the property

The certificate will confirm:

- That the contractor who undertook the underpinning works was, to the best of the Expert's knowledge, competent to undertake the work
- That the Expert has used professional skill and care to design the works and also that Building Regulation approvals were obtained for the work (copy documents to be given with the certificate)
- That where the contractor designed the underpinning works, the Expert had reviewed the adequacy of the design
- That the Expert monitored the works during the course of the contract and was satisfied with observations at the times of the visits

The Owner should be provided with a final set of 'as built' drawings and a copy of the specification along with the Certificate of Structural Adequacy, as these should record and confirm the extent and nature of the remedial works undertaken to the foundations. This does not diminish the legal rights of the Owner to pursue any party for defective work. The issue of the Certificate of Structural Adequacy should be the last step in the contract after all works have been completed and all amounts agreed, certified and paid.

11 Drivers for change and industry innovations

11.1 Introduction

Subsidence is one of the most costly ground hazards in the UK, with thousands of claims made each year for damage to properties from differential ground movement. When summers are hot and dry, large increases in claims are observed, especially where shrinkable soils are present. These patterns and trends are well known to subsidence Surveyors and claims handlers, and increasingly are also of interest to professionals in other industries.

Over the past 20 years, there has been a considerable increase in the use of data and analytics in insurance and mortgage lending, and the subsidence industry has itself become increasingly digital and data intensive in nature. In addition, the scope of interest is expanding; from individual subsidence cases to establishing and refining regional and increasingly local patterns across space and time. These trends are set to continue.

Increasing focus is being brought to the creation of robust, data driven subsidence maps and models which can identify areas of the country at increased risk from ground movement. In addition, as our climate continues to change to one with warmer, wetter winters and hotter, drier summers (Figure 11.1) we can expect increases in the number of properties at risk from ground movement. As a result, more attention is being paid by the financial sector to the impact of climate change on subsidence.

The aim of this chapter is to outline some of the changes in the industry's view on how to model, map, manage, measure and mitigate the impact of subsidence on properties, while at the same time improving the outcomes for clients (Homeowners), Insurers and the environment.

Figure 11.1: Changing mean UK monthly temperatures (1981–2010 and 2010–2020 baselines) and UKCP18 RCP 4.5 median scenario (2050s and 2080s)

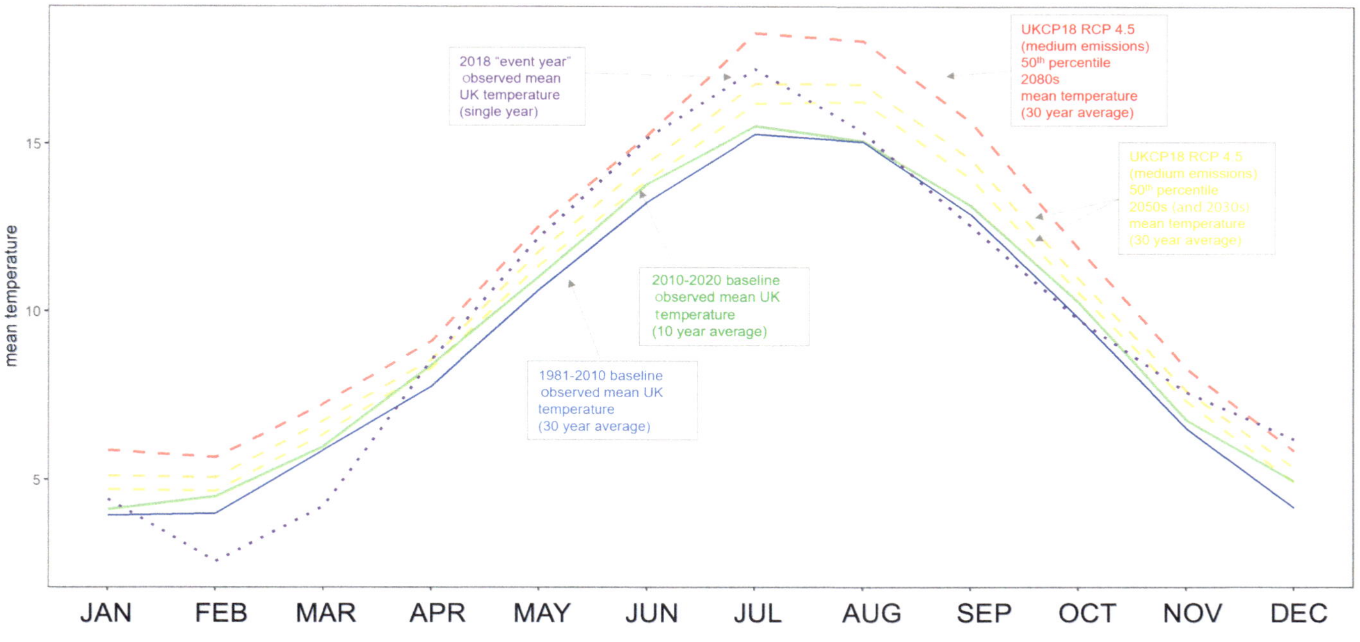

11.2 Changing needs of the subsidence industry, regulators and Homeowners

Like many other sectors, the subsidence industry is facing a number of different drivers for change. In the face of the climate crisis, there are strong ethical, environmental, commercial and regulatory voices which demand rapid and significant reductions in the carbon and environmental costs of remediating and preventing subsidence. Alongside these required changes, there is a growing unease in the industry both at the length of time that subsidence claims can take to be finalised, and also a recognition of the impact that these protracted negotiations and resolutions can have on the mental health of Owners. There is also an increasing appreciation of the many benefits of tree canopy cover, green spaces, and permeable surfaces within the urban environment, especially as these areas will need to be adapted to cope with hotter, drier summers and more extreme weather events.

It is clear that the decades-old remediation techniques (e.g., carbon heavy concrete underpinning, tree removal etc.,) need to be accompanied by or even replaced with improved stabilisation approaches that allow for trees in close proximity to houses. Tree-foundation interactions will become an increasingly common problem as the average garden size continues to decrease, and the requirement for trees in the urban environment will continue to increase. We know trees reduce city temperatures as well as provide numerous health benefits. There is an urgent requirement for governments and regulators now to shift the whole construction industry towards low-carbon foundations designed for close interaction with large vegetation.

The Bank of England is now requiring Lenders to assess the impact of climate change on their property portfolios. Even insurance companies are requesting access to climate-adjusted subsidence models. Thankfully, a community of engaged scientists and businesses are developing new ways to address many of these concerns.

11.3 Changing approaches to measuring and modelling subsidence risk now and in the future

Better analytical modelling using greater amounts of accurate and relevant data will lead to more effective and accurate prediction of subsidence in terms of: numbers and likely location of claims; effective prevention; cost effective claims management and improved client satisfaction.

For over 20 years, UK insurance companies have licensed or created subsidence models to help them identify properties at increased risk. Two distinct approaches dominated the modelling for a number of years. One was based primarily on clustering previous claims exposure, and the other based on mapping the shrinkability and related properties of the underlying soils or geology. More recent models integrate both approaches, and over time more factors have been added to these models, including weather patterns and tree locations.

As previous chapters describe, subsidence damage to a property is often a highly site specific and complex process, and typically involves multiple interrelated factors.

These include:

- Soils. The ability of the soil to shrink as it loses water (clays), or washout (sands and silts) or compress or shrink (peats etc.,) and the highly local changes in soil texture
- Weather conditions. Typically temperature, rainfall and evapotranspiration, often summarised in derived metrics such as 'soil moisture deficit' (SMD) (Figure 11.2)
- Trees. Location, size and species of trees and their drying effect on clay soil strata
- Property construction. The age and fabric of the building and the materials from which it is constructed
- Variable foundation depths. The shape of the building, the foundation depth and whether there is a cellar or deeper foundations under a later extension. Deeper foundations under part of the building can act as natural pivots on which the remainder of the building hinges, or can lead to differential settlement
- Drains. Location and age of drains, together with their proximity to the foundations. Where the ground is especially silty or sandy, leaking drains can soften or wash away the substrata leading to a loss of strength in the underlying soil

Figure 11.2: Predicted August soil moisture deficit (SMD) for the 1981–2010 baseline and various periods under the RCP 8.5 scenario (high emissions)

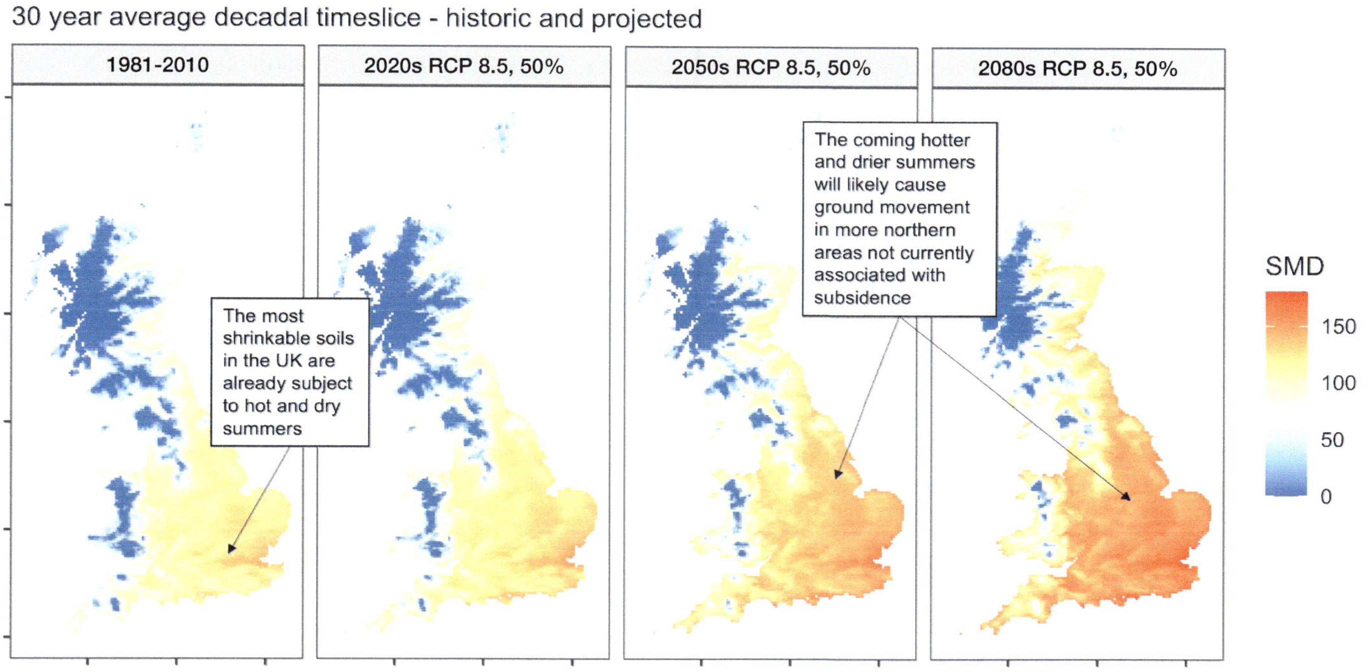

All of these factors are capable of being 'digitised' i.e., represented by data suitable for input into a predictive model. Most in-field subsidence Experts already intuitively take most or all of these items into account as the cause of damage is diagnosed and the best course of action to take is established — yet the direction of travel for the industry is to increasingly seek to understand likely process off-site.

Going forward it is highly likely that all of these factors will become more integrated in algorithms used by Insurers and Lenders. Historically, the complexity of these models prevented manual calculations to establish the relative impact of the contributory factors leading to subsidence. However, with the advent of environmental data science, multi-factor models are now commonplace. As machine learning develops, and environmental and claims data becomes more available, usage will increase beyond the insurance and mortgage sectors and may even form a feedback system, with in-field Surveyors and Assessors contributing to the improvement of these models though the contribution of verified and accurate in-field evidence. This is a key point, as a model can only ever be as good as the data that goes into it. As a result, most of the effort in building subsidence models comes from cleaning and processing the data, long before the modelling begins.

Some traditionalists may still consider the diagnosis of subsidence as being closer to an 'art' than a science, and as a result an analytical approach is likely to remain contentious for some. Nevertheless, such models are already in use by a number of companies in the industry, and the direction of travel is clear.

The impact of the increased use of data and analytics will be that the likelihood of the mechanism of subsidence damage will be better understood, and that off-site risk calculations will become more precise and granular when compared to the relatively broad-brush approach used for the last few decades. The accuracy of underwriting pricing is already improving. Similarly, the Bank of England has required all large Lenders to consider the impact of climate change on subsidence rates for their property portfolio. Numerous banks are even using these forward-looking models to inform their lending processes, identifying properties at increasing risk from subsidence over the next 30 years. A similar approach is already being used to alert Owners to a potential problem — with new reports available to alert Homebuyers to future risks to their prospective properties. These reports align with the desire of Insurers and Lenders to be more proactive when it comes to environmental risks.

As Insurers, Loss Adjusters and Surveyors have become more open with sharing GDPR-compliant versions of their historic claims data, it has become possible to develop a new suite of subsidence models which combine the physical properties of the ground, local weather conditions, tree data and historic claims. In addition, some of the new models also consider other hidden ground risks, such as the presence of historic mine workings, running sands, natural voids or karstic environments, landslide-prone areas etc. As such, these models are becoming more useful for claims handlers, who can determine likely dominant causes of ground movements even before a site visit. New models will include more property attributes, such as the construction type and age of the property.

As well as considering ground conditions and historic claims, novel remote sensing approaches are being developed to monitor the movement of the ground. An increasing number of companies are offering radar satellite-based data products (InSAR) to detect ground movements, and these are showing signs of promise, especially in areas where there is continual movement in one direction, e.g., when a new tunnel is constructed under a city. While there is still much development required and scope for cheaper, higher resolution sensors, in time it is hoped that these technologies will be able to detect the small-scale seasonal differential movements which lead to the majority of subsidence issues in low-rise buildings.

As our climate changes, the old approach of Insurers simply looking at historic clusters of subsidence claims at postcode level will become increasingly untenable. Already, climate-adjusted models are available which train models based on observed historic weather patterns and claims, and then look forward to apply this knowledge to the weather conditions expected for the coming decades. Such models are already in use by major Lenders, as required by the Bank of England. Similarly, this information is now regularly available as part of the conveyancing process, enabling Owners to understand the changing risk of subsidence to their property through time.

Related work predicting the impact of climate change on other assets in the built environment is ongoing, with infrastructure companies also keen to understand the longer-term environmental risks to their networks. In all these areas the collection of clean data on the date and location of the failure or damage is incredibly important, as the more we understand about the timing and location of these events, the better we will be able to predict and prepare for the coming impacts.

11.4 Changing technologies to reduce the financial, time and environmental costs of claims

Digitalisation will improve underwriting in terms of cost and accuracy; taking into account the likely ongoing impact of climate change, and allowing Insurers, intermediaries and Owners to reduce their carbon footprint throughout the claims process.

Both insurance companies and Owners are keen to avoid the financial, environmental, time and emotional costs of addressing damage from subsidence. Lengthy subsidence claims trap people in their properties and can cause significant impacts on mental health and financial security. Across the industry, new approaches to reducing the time taken to address claims, as well as the environmental impact of the claim, are being introduced.

A number of technological innovations are being employed to improve the handling and resolution of claims. This section provides some insights as to current and likely future uses of novel technology going forward, and how this may impact on processes and the necessary skills of the Expert. It is not intended to provide a comprehensive guide to the technology involved but rather to raise awareness which might invite the reader to carry out further technical research.

There are a series of 'mega trends' which the reader should be aware of:

- The increased movement towards digital products and services, providing data which allows both greater insight and actionable outcomes i.e., the likelihood that existing processes might be reviewed and possibly changed. The reality is that this may not be simply a matter of digitising existing processes but rather that entire new approaches may start to emerge
- Greater availability of more open environmental and claims data
- Increased (and cheaper) computing power will improve the accuracy of the prediction of subsidence and transform underwriting. This may prove to be especially important as the impact of climate change not only shifts the areas

which are most vulnerable to subsidence further north, but may also affect the behaviour of trees which are viewed as being one of the principal causes of clay-related movement
- The use of advanced analytics and ultimately AI in some form which will help Experts make more sense of their data
- An increase in the use of social and other digital media which not only transforms the way the Owner interacts with the Insurer but potentially with the entire supply chain
- Decarbonisation of the industry and prioritisation of urban ecosystems goods and services

As many claims span two or more years, and require a number of Experts to attend the site, the distance travelled to simply inspect the damage to the property can be significant. There have been emission reductions achieved through an increase in the number of surveyors using lower emission, hybrid and now electric vehicles, and we expect this trend to extend to contractors and builders in future.

Alongside this, there is an ambition to continue to reduce the number of site visits required through the increasing use of electronic communication and digital evidence collection. As discussed in Section 11.3, some initial triaging of subsidence claims can be undertaken even before site visits are begun through the use of new subsidence and ground risk models. In addition, some companies now offer secure evidential video and photograph collection, enabling Owners to upload unedited evidence of property damage. Digital claims management can also increase the speed of decision making, and reduce costs and uncertainty for all parties. Furthermore, through digital claims management, the increasing number of interested parties can access the relevant information at the same time.

Increased digitisation and improved communication methods will allow the Owner to 'look into' their case file with greater transparency. The effect of this will be that the Owner will no longer need to contact the Insurer or their representative to seek an update but rather that the operating model might increasingly move to a form of 'self-service' governed by a set of automated rules, and one where the Owner can play an active part in scheduling activities.

The main impacts of this will not only be an improvement in customer service as a result of greater flexibility and transparency but also a reduction in the operational cost by the Insurer. Additionally, there will be greater consistency of approach which will assist not only in benchmarking and the audit process, but will lend itself to yet further improvements in underwriting accuracy and claims handling.

The impact of the Covid-19 pandemic in terms of being unable to carry out physical inspections has led to an acceleration in the use of remote imagery and virtualisation. Both of these types of 'unstructured' data, when coupled with 'virtual site investigation' have opened the door to more remote methodologies which may ultimately affect the way that subsidence is managed, both at a risk and claims management level. Although the industry is some distance away from the use of virtual and augmented reality, we see claims handlers embracing this technology for other losses and it is quite likely that this trend will continue.

Insurers are keen to decrease the environmental damage resulting from their actions to remediate subsidence damage, to the extent that some are already considering the carbon costs of remediation options. Tree removal, the various forms of underpinning, and novel (controversial) approaches such as resin injection are available options. But which is the 'best' choice that stabilises the property for a fair cost and causes the least environmental damage?

More work is required to develop efficient and effective approaches to determining the most environmentally sensitive options to repair subsidence damage. For example, when considering whether to remove a tree or underpin a house with concrete, there is a tension between the loss of sequestered carbon and a range of ecosystem goods and services provided by the tree, and the carbon and wider environmental costs of the concrete. Both are damaging, but which is the 'least-bad' choice? Perhaps in such situations there is scope to develop a new damage remediation model which brings together all the relevant Experts at the same time to both triage claims and, where possible, determine the fastest and most effective course of action to enable the claim to be resolved.

11.5 Changing research areas

Continued focus on relevant research into trees and soils, aided by new technologies will, through greater collaboration, lead to greater understanding into the mechanisms of subsidence movement and subsidence-resistant foundation design.

In such a broad subject area, there will always be a number of key areas which could benefit from further research and development. The following subsections discuss just a few areas which could help reduce the environmental impact, time and frustration associated with remediating subsidence claims.

11.5.1 Trees – increasing coverage and benefits while reducing risks

More research is required to determine why some trees seem to be more associated with subsidence claims than others, even within the same species. This will require consideration of tree biology, local soil conditions and the subsurface urban environment (e.g., sewers, local hydrogeology).

In the same area, more work is required to encourage the safe 'greening' of the urban environment. With this in mind, a selection of trees which can be safely planted on the most shrinkable clay soils should be identified and publicised, with the planting of such trees encouraged in place of felled trees.

Likewise, more enhanced approaches to quantifying ecosystem goods and services are required, in order for non-specialists to undertake more holistic assessments as the best course of remedial action. Similarly, better quantification and communication is required of the efficacy of tree options such as the installation of root barriers, pruning approaches, and felling. Such research could demonstrate the success of these interventions, by tree species, in returning a property to stability.

11.5.2 Urban soils, soil ecosystem health and the interface between soils, vegetation and construction materials

There is an urgent need for enhanced mapping of the soil subsurface in urban environments. Surveyors can contribute to this through the sharing of results from soil samples. The soil, even in an urban environment, is a complex ecosystem in itself, in which the vast majority of life is invisible to the naked eye, and these ecosystems require protection to allow them to continue to provide their valuable services.

Subsidence repair will often involve underpinning which results in a variety of construction materials coming into contact with soils and vegetation, most notably tree roots. Pumped or pressure injected materials may be cementitious, chemical or resin based. In view of the real need to ensure the continued good health of soils, trees and vegetation, particularly in the urban environment, research needs to be carried out into the various soil/construction material interactions.

While pressure injection may appear to be an innovative and cost-effective remediation tool, more research is required on the impact of pressure injection approaches to soil health, and provision of ecosystem goods and services, especially when a significant volume of pumped material is injected into the soil. Likewise, any impacts on vegetation, soil, fungal, animal or human health should be identified and quantified, before use of injected material becomes more widespread. Records should be kept where novel systems are used so that long-term impacts to trees and large vegetation can be monitored.

11.5.3 Mobile sensor and in-field cause determination

To increase the speed of cause determination, there is significant scope to enable more in-field decision making. While we may be some way off in-field root analysis through mobile light microscopy or DNA species identification, the pace of mobile phone development, image recognition and AI raises the possibility for future advances in these areas. Likewise, with image recognition apps becoming better at tree species identification, there is scope to develop a suite of AI-assisted in-field cause-determination tools. Apps have already been developed to instantly assess certain soil parameters, and this is an area ripe for innovation.

11.5.4 Remote sensing of ground movements

There is excitement from a number of quarters about the possibilities of the use of remotely sensed data to track the seasonal movements of soil and the built environment. Radar-based satellites such as the Sentinel satellites offer a new approach (InSAR) to monitor near-real-time movements of ground.

However due to the size of the pixels for these radar satellites, the small-scale movements responsible for most subsidence cases are often not identified. This makes it difficult to incorporate large-scale interferometric observations of movement into helpful subsidence models at present. One of the other issues with this type of remote sensing for the purpose of confirming movements is that it picks up on other movements, for example thermal expansion of buildings in the summer months. Such movements can confuse or obscure other signals.

More work is required to improve the capabilities of satellite technologies, such as InSAR, to measure and even predict subsidence damage. Current challenges exist with high levels of false positives and false negatives, but hopefully with the advent of higher resolution sensors in the coming decades, and enhanced algorithms, observations of differential movement on a single property (even accounting for thermal expansion) will one day be achievable. Such capabilities would enable near-real-time monitoring of property level movements.

Aligned with this, further improvements in the high-resolution mapping of urban soils are always welcome. The development of improved tree-species maps from remotely sensed hyperspectral images will also enhance the ability of all parties to determine potential causality more quickly and efficiently.

11.5.5 Low carbon, tree-friendly foundation design

Finally, there is a pressing requirement for the establishment of a new industry-standard, low (or zero) carbon foundation for low-rise buildings, designed to tolerate coexistence with large tree roots on shrinkable soils in hot and dry conditions. Current standard foundations use vast amounts of carbon-emitting concrete (contributing to climate change), which are not sustainable in a low carbon economy. Some modern foundation designs can enable planting of trees in closer proximity to houses, enabling significant greening of the urban environment and unlocking the many associated benefits — including net reductions in carbon and reductions in extreme temperature, as well as improving air quality and human physical and mental health.

Lower carbon foundation designs are currently 10–15% more expensive than traditional trench fill foundations, but such small cost differences could be reduced or eliminated if they became the standard foundation and benefitted from economies of scale. Discussions with home builders have indicated they are ready to make the change to low-carbon housing, if the government forces all builders to embrace this approach. Thus, for the industry to adopt low carbon foundations, urgent political action is required to mandate change.

11.6 Cross-sector engagement

Engaging with academics, businesses and other specialists outside the subsidence community will enable more rapid progress in addressing these exciting areas for improvement. Increased sharing of subsidence data, tools, and learning can enable the construction, insurance and remediation industries to rise to meet these challenges. With firm commitments from governments and industry to leave the world better than we found it (and robust legal frameworks to support these commitments), we can improve the structural and environmental health of our villages, towns and cities, creating a more resilient built environment that can meet the challenges of global climate change.

Appendix A Causes, mitigation and prevention

A1 Introduction

The majority of this publication deals with remedial measures. This appendix considers what steps are needed to mitigate damage — or to prevent damage from occuring altogether.

As a basic tenet, insurers are concerned with the effect of damage that has occurred, rather than preventing damage that may occur. They view prevention as a normal duty of care by the Homeowner and something that should be carried out at the Homeowner's expense. Insurers are not concerned with the cost of work needed to prevent damage occurring and will sometimes view a lack of action by the Homeowner as a breach of the conditions of the policy. In reality, they normally take a pragmatic and sensible approach to what might be reasonably foreseen.

Where damage has already occurred, the insurance contract allows for the cost of repair but wider considerations take into account the possibility of future damage. The 'Public Authorities Clause' of most insurance policies give the Insured added protection in funding the cost of avoiding future damage, because of the need to comply with Building Regulations.

Before preventative or mitigation solutions can be developed, it is important to understand the cause (or likely possible cause) of the subsidence problem.

Where damage has already occurred, thorough investigations are required following the guidance given in Chapters 5–7. Once the problems are understood, various mitigation options should be considered, singly or in combination:

- Stabilise the subsiding ground under the foundations (ground treatment)
- Transfer building foundation loads to deeper, competent soil strata and isolate volume-changing soils from the foundations — various underpinning techniques
- Stabilise soil slopes — slope stabilisation
- Stabilise/strengthen existing retaining walls

The following sections are a route map to successful environmentally friendly preventative or stabilisation solutions, cross-referenced to the detailed guidance provided in this and other guidance.

A2 Potential causes of foundation subsidence

More detail can be found in Chapters 5 and 8, but causes of foundation subsidence could include:

- Tree root influence
- Leaking drains
- Leaking water mains
- Chemical attack of concrete
- Collapse of subterranean mine workings and subsequent progressive subsidence of upper soils into the mine workings
- Progressive subsidence of upper soils into rock dissolution features
- Influence of excavations nearby (fine-grained soils changing from undrained to drained soil states or, in granular soils — a loss of support)
- Construction of underground tunnels under or near to existing building structures
- Water table changes (burst water mains, new drainage ditches, new basements nearby, climate change effects, groundwater extraction)

- Frost heave in frost-susceptible soils
- Vibration from road traffic, driven piling construction or activities, reciprocating machinery, dynamic ground improvement techniques (causing compaction or liquefaction of the soil or resonance in the structure)
- Erosion (water, wind and freeze/thaw actions)
- Volumetric changes to soil (shrinkage and heave) — tree growth and tree removal, evaporation due to the effect of direct sunlight, and less evaporation in shaded areas
- Lime/cement stabilised soils swelling due to high soil/water sulphate content
- Inadequate foundation design due to insufficient knowledge of the ground conditions
- A failure to calculate the magnitude and disposition of loads on the foundation
- Location of compressible material like peat or expansive materials e.g., sulphates and sulphides beneath foundations
- Defective construction of the foundation

The scope and diversity of causes of subsidence inevitably means that damage cannot always be anticipated. Some causes such as the impact of trees on clay are perhaps more apparent, but even then there is considerable uncertainty.

It is very unusual for Insurers to decline to deal with a claim for subsidence damage unless it is evident that there has been a failure to take appropriate preventative action.

A3 Options to mitigate the effects

Mitigation measures could include:

- Tree management (Chapter 8)
- Repair drains (Section 9.2.2)
- Repair leaking water mains (Section 9.2.2)
- Neutralise chemical attack
- Remediate subterranean mine workings and rock dissolution features — specialist pressure injection grouting, on a close grid, plus validation report[97, 98].
- Backfill adjacent excavations with engineered fill, or transfer existing building loads to deeper soil strata — underpin existing foundations adjacent to the excavation
- Design and construct retaining walls to stop the existing building subsiding into the excavation nearby
- Design and construct remedial foundation repair solutions to eliminate the effects of frost heave on frost-susceptible soils — dig out frost-susceptible soils around the perimeter of the building and replace with non-frost susceptible soils using a ‘hit and miss’ construction technique. Alternatively, consider geopolymer injection to change the properties of the soil to non-frost-susceptible
- Design and construct remedial foundation repair solutions to minimise the effects of vibration (Section A7)
- Design and construct remedial foundation repair solutions to minimise the effects of erosion
- In conjunction with tree management techniques, design and construct foundation strengthening schemes to minimise the effects of volumetric changes to soil (Sections A4–A6)
- Where an existing foundation construction is proved to be defective and/or the design deficient, design and construct a foundation strengthening scheme to mitigate the problem. Ideally, this solution should be environmentally friendly (with low embodied carbon and minimising damage to trees)

Of these, the Homeowner is likely to be aware of (and therefore in a position to take mitigating action against) the first three only.

A4 Eco-friendly mitigation

Wherever possible, it is important to use eco-friendly ground improvement techniques to mitigate the effects of subsiding ground under existing buildings — in both shrink or heave scenarios.

Wherever possible, existing trees should be protected and not disturbed. Amenity value of trees should be considered.

Any remedial works to existing foundations should be engineered to avoid damage to trees where practically possible.

Ideally, all remedial subsidence mitigation solutions should minimise the amount of soil leaving site. Remedial solutions with low embodied carbon should be given due consideration and duly weighted with other factors like cost etc.

As an alternative to removal, the existing soils under the foundations can be engineered and reinforced, using appropriate techniques, to provide adequate soil strength and substantially reduce volume change potential. Where this is not possible, eco-friendly underpinning options should be considered and adopted where feasible (Section A6.3).

A5 Soil improvement (to reduce the effects of heave)

Subsidence may be avoided by improving the soil. This is particularly relevant in the case of new build construction, on sites where trees have previously existed.

In a fine-grained soil with a medium to very high plasticity index, the soil acts as a natural hydraulic jack. If it is desiccated and subsequently absorbs moisture, it expands/heaves. The soil can heave both vertically and horizontally.

If this soil is shrinking due to the water demands from external influences (probably due to tree root soil moisture extraction), it transforms into a pseudo stiff semi-cemented state and reduces in volume.

Unconfined shrinkable clay with heave potential is not usually a problem — e.g., in soft landscaped garden areas. However, problems do occur in hard landscaped areas.

If expanding volume change soil is able to push against an existing foundation, it can apply significant upward and horizontal forces on the foundation — typically more than 100kpa of upward pressure. This clay heave uplift pressure is sufficient to lift a building foundation, within the clay rehydration area. Upward soil swelling movements can be up to 150mm in high volume change potential clay, up to 100mm in medium volume change potential clay and up to 50mm in low volume change potential clay.

A6 Methods for preventing further subsidence

A6.1 Soil moisture content control systems in clay soils

Control of ground moisture content would seem, at face value, to be an effective way of preventing or mitigating damage. Such an idea is not without its problems however, given the complex nature and properties of clay as a material.

Although not currently available, but as a possible vision for the future, imagine installing a system of soil moisture content sensors under and around an existing foundation, connected to a remote monitoring station (well points with sensors at various depths, with irrigation pipes integrated into the well point — a porous sealed unit).

The remote automatic monitoring station would control an accessible and maintainable site water treatment manifold connected to micro water injection/extraction well points.

This manifold would be connected to a solar powered pump, with battery backup, capable of micro pumping water into the ground, at various depths (seepage treatment using several multi-holed irrigation pipes per well point) from a grey water supply storage tank. The system would also be capable of micro-extracting water to the grey water storage tank if required.

Before the system is installed, the fine-grained soils (clays) would be tested to determine their equilibrium moisture contents (the moisture content value where the soils neither heave nor shrink).

The monitoring station would be controlled by software that interprets the soil moisture content data and then controls the treatment of the ground — either multi-level water injection or extraction. The restricted access equipment required to install these well points would require development, maybe using a driven installation or a micro auger technique? The system would likely be activated once a week and could be triggered by new 'smart home' technology.

Existing trees would remain and damage to tree roots during the installation of the system would be minimal. Minimal soil would be removed from site — just the soil displaced by the well points — but this would depend on the method of installing the points.

A6.2 *In situ* ground improvement of shallow soils

Where a predominantly granular soil under a foundation has been loosened by the action of water, frost, adjacent excavations or vibration, the strength of the soil under the foundation could be increased by using a ground improvement technique. This is particularly relevant to infill i.e., 'brownfield' sites.

Where practical, it is important to rectify the cause of the soil-loosening prior to ground improvement.

Appropriate ground improvement techniques could be:

- Geopolymer injection of the soil supporting the foundations
- Hit and miss underpinning using soil-mixing additives to improve the strength of the excavated soils (excavate the soil, improve its strength by soil-mixing with other binders and then put it back in compacted layers and possibly use temporary flat jacks to transfer the load from the existing foundation to the improved soil, and then semi-dry pack)
- Hit and miss mini vibro compaction treatment using a guided vibrating poker or grundomat. This technique would require the importation of suitable granular fill to the site to compensate for the densification of the treated soil. Currently, such a solution is not available but as the pressure on infill housing continues in the UK, there may be scope for further innovation in this area

A6.3 Underpinning

Wherever possible, from an environmental point of view, underpinning techniques should minimise the amount of soil removed from site and should minimise the amount of concrete used.

Typically, traditional methods of underpinning have a high embodied carbon content (large quantities of soil removed from site to landfill tips and large quantities of concrete imported to the site to construct the remedial works foundation solution).

Usually, traditional methods of underpinning comprise of one or more of the following:

- Hit and miss mass concrete underpinning
- Piled raft underpinning incorporating under-slab void formers
- Hammerhead piled RC ring beams
- Stem base underpinning — where soil retention is required

All these techniques generate soil waste product, use concrete and can cause significant damage to tree roots. New eco-friendly underpinning techniques are required and ideally, the materials used to underpin a building should be recyclable.

Where piling is considered viable, it may be possible to strengthen the walls of an existing building to span to new, improved pile caps (introduce bed joint reinforcement to masonry walls to create beams). The system would be engineered to limit bending, shear and bearing stresses to acceptable limits compatible with the masonry material properties.

The piles could be made from recycled materials (thereby, extending the original life and use of a material that would otherwise have been discarded or melted down etc). Old piles could even be reclaimed from the ground.

A7 Minimising subsidence caused by vibration

This section considers how vibration-induced subsidence can be avoided or reduced. It is caused by:

- Heavy vehicles driving over a pothole or a ridge in the road surface. The wheels of the vehicle can transmit a vertical cyclical dynamic force into the ground and then propagate a compression wave into the soil. Sands and gravels are good at transmitting compression waves
- Road traffic noise

- Nearby construction activities such as pile driving, vibro compaction and the compaction of engineered fill
- Reciprocating machinery (e.g., heavy metal presses and guillotines)

If loose granular soils are supporting an existing building foundation, prolonged vibration of the soil can cause the soil to densify with corresponding decreases in volume. The foundation drops in proportion to the decreased volume of the soil, potentially leading to defects in the superstructure.

Remediation of the problem could be a combination of the following:

- A clay slurry cut-off wall installed on the road/property boundary, into the ground to sufficient depth. The amplitude of the roadside compression waves would be reduced by the damping effect of the slurry/soft clay wall. The cut-off wall could have an automatic irrigation system installed to prevent it drying-out
- Geopolymer injection treatment of the loose granular soils under the existing foundations
- Other underpinning techniques
- Installation of vibration isolation bearings between existing foundations and the superstructure

A8 Stabilising soil slopes

Soil stability is a complex geological phenomena and can be especially problematic where it leads to property damage. As a result, it is important to investigate and establish the cause/s of soil slope instability before remedial works are designed[99].

Sometimes, where a soil slope is large and moving downhill, it may not be practical or economically viable to stop the movement. The sliding forces may be too large to be arrested and stopped by the design and installation of remedial engineered systems such as piled retaining walls, soil mixing techniques to form retaining walls, soil nails, ground anchors, gabions, reinforced earth, drainage systems etc.

The implication is that a property may sometimes be beyond economic repair.

Where predicted long-term slope or ground movements are small, it is possible to underpin a detached building by stooling up the building, by installing a groundbearing raft with jacking points in it and then casting an RC slab on top.

In this approach, the upper RC slab has the building sitting on it. The groundbearing raft is connected to the upper slab only by the jacks. The jacks are suitably connected to the slabs to ensure adequate horizontal stability. A synchronised jacking system can be designed to periodically jack the building back to level. Incoming and outgoing services must be designed to accommodate horizontal and vertical movements over the service life of the building. In essence, the building rides on the moving slope. It is a relatively expensive solution with a high embodied carbon content.

The impact of this technical solution on the value of the affected property should not be underestimated.

Where it is practical to stabilise a soil slope, there are numerous options[100]:

- Embedded solutions — concrete pile retaining walls and steel pile retaining walls (cantilevers and tied cantilevers)
- Gravity solutions — RC retaining walls, stone-filled gabion retaining walls, interlocking dry-laid masonry or reconstituted stone retaining walls, reinforced soil-retaining walls, timber or concrete criblock retaining walls
- Soil mixing techniques to form a contiguous wall and hence act as a retaining structure
- Reinforced or nailed slopes
- Improve soil pore water pressure management by the introduction of appropriate drainage systems within the slope
- Plant trees to reinforce the slope (tree roots holding the soil together) and to lower pore water pressures in the slope (water extraction by tree growth). Trees are heavy, so it is important not to plant them on areas of the slope where the extra weight could reduce factors of safety for sliding and slip resistance[101].

The remedial works from a structural point of view would principally comprise two stages:

- Stage 1 — remediate movement of the slope
- Stage 2 — stabilise the subsiding foundation

Appendix B Certificate of Structural Adequacy

This appendix contains a template for a 'Certificate of Structural Adequacy'. Note, it is not a prescribed format and Experts should adapt the document to suit individual circumstances.

Its purpose is to provide additional peace of mind to the client (usually the Owner, but possibly the Insurer/Lender if the certificate is to be used for such purposes).

The template should act as a prompt for items that should be included in/removed from the final certificate.

The following general points should be considered prior to issue:

- The Expert should obtain approval from their PI Insurer with regard to the wording of the certificate
- Terms of Appointment should always be in writing and clearly define the responsibilities the Expert undertakes with regard to inspection and certification of the works. If (perhaps in order to save fees) the terms curtail the services which the Expert believes are required in order to issue a certificate with confidence, they should inform the client immediately
- All site visits by the Expert during the course of the project should be recorded (and a record kept of the general state of the works at the time of the visit)
- The Expert should delete (rather than leave blank) any part of the certificate which is not applicable. This may require prior discussion with the PI Insurer. For example, for a project where a property has been previously underpinned, but for which the Expert was not involved, and the 'report' option is being followed as opposed to the 'works' option, the name of the Contractor who undertook the original work may not be known. Alternatively, if drawings and specifications are not available for the works then the extent of the works may not be known. Any limitations as to the professional assurance given should be made clear in the endorsements section
- The underpinning contractor's guarantee should be for a defined period (usually 10 years) and should be insurance backed. The Association of Specialist Underpinning Contractors has a relevant scheme in place (Appendix F). The certificate should refer to the guarantee and ideally include a copy for ease of reference
- Reference is made to the Expert's 'report'. The Expert should refer to IStructE guidance on the preparation of reports relevant to surveys and inspections of buildings and similar structures[102]
- If other Experts have provided advice during the course of the project, this should be noted on the certificate

Specific aspects:

- The certificate is not prescriptive and should be adapted as appropriate; the spaces provide are not indicative of the level of suggested detail
- **Project details:** This should be self-explanatory
- **Damage and repairs:** This section should be used to summarise the damage that had occurred, the cause, and the proposed repairs — including reference to the drawings and who carried out the work
- **Design assessment:** If design work was carried out by a specialist (e.g., a piling contractor), the details should be recorded in this section
- **Inspection of repairs:** This section should record the site visits and monitoring that has occurred as part of the works. It should make an assessment of the repairs carried out
- **Endorsements:** If appropriate, this section can be used to make comments about aspects the Expert could not see or assess
- **Revisions:** It is good practice to note any necessary changes to the project
- **Signatures:** This should be adapted to suited company practices

Certificate of Structural Adequacy (template)

Company Name, Address, Contact Details, Company No.

YOUR BUSINESS

Project details

Client:
Project address:
Terms of appointment:
Project reference:

Damage and repairs

Damage to property:
Causes of damage:
Proposed repairs:
Report reference and date:
Drawings and design details:
Contractor:

Design assessment

Design undertaken by:
Design reviewed by:
Design document reference and date:
Assessment of the design:

Inspection of repairs

Site visit date and details:
Site visit date and details:
Site visit date and details:
Assessment of repairs:

Underpinning details

Underpinning contractor:
Installation dates:
Guarantee reference, date and length:

Endorsements

Certificate

This certificate should be read in conjunction with **Insert any relevant documentation here**
This certificate is a considered professional opinion and is not a warranty or guarantee as regards the works undertaken, and no liability shall attach to us except to the extent that we have failed to exercise reasonable skill, care and diligence in the provision of our services under our Terms of Appointment. This certificate does not in any way increase our liability beyond that agreed under the Terms of Appointment. This certificate is not a certificate as defined in The Building Act 1984 and The Building (Approved Inspectors, etc.) Regulations.
This certificate is for the sole benefit of the above-named client and may not be relied on by any other person whatsoever. Transfer of the benefit of this certificate to any party other than the client to whom it is addressed will only be permitted with our written permission. Before agreeing to any transfer of benefit we may require to re-inspect the property, for which a fee may be charged.

Revisions

Signatures	Prepared by	Checked by	Approved by
Name			
Job title, post-nominals			
Signed:			
Date:			

Appendix C Party walls and neighbouring properties

C1 Introduction

The nature of repair works to existing properties is such that an Owner must accept the complexity of the surrounding built environment — which sometimes requires interaction with neighbours. In some cases there are strict legal procedures which need to be followed. In other cases, it is a matter of practicality and the application of common sense, aided by effective communication and cooperation.

Some repairs cannot be achieved without the access and facilitation of the neighbour. In addition, some repair works while resolving a problem for one Owner places the neighbour's property at risk of damage.

This appendix is limited in scope and expressly excludes new build, and full redevelopment construction works. The legal system to facilitate repairs to existing properties in the UK has evolved such that the express right of a neighbour to refuse access is overridden, subject to the correct administration of the relevant processes explained in Section C2.

C2 Access by neighbour consent

Relations between neighbours are usually amicable, and most neighbours are understanding to the plight of an Owner needing to repair their property, seeking reasonable access.

However, it is often appropriate for a neighbour, when asked to provide anything beyond the most basic visual inspection access, to also ask for the comfort and protection of a written agreement known as an 'access licence'.

The broad nature of access requests (from a few hours on a single day, to prolonged inconvenience that could last for several months) means that access licences can be as simple as a standard template agreement — issued by the Owner's consultant with basic terms and typically including a schedule of condition to allow before and after assessment of damage — to the most challenging; requiring both parties to appoint Solicitors.

It is perfectly reasonable for a neighbour to request that access must be documented in a licence. This is not a rejection or a dispute and should not be considered as unneighbourly.

C3 Access to Neighbouring Land Act 1992

Until the Access to Neighbouring Land Act 1992[72] came into effect, Owners of properties in disrepair were at the mercy of their neighbours, who could reject perfectly reasonable access requests. The Act enables the Owner to carry out those reasonable works provided they are for the purpose of the basic preservation of that land. In the context of subsidence, this wide scope is important as the facilitation goes beyond works to the building — covering access for the purpose of helping to maintain trees and hedges, and any necessary work to drains.

If a request is refused by the neighbour (and is subsequently found to be reasonable by the Court) the Court can grant the Owner an 'access order'. The Court must be satisfied that the following two conditions under Section 1(2) of the Act have been met:

(a) That the works are reasonably necessary for the preservation of the whole or any part of the dominant land.
(b) That they cannot be carried out, or would be substantially more difficult to carry out, without entry on the servient land.

It will therefore be necessary for the consultant team to have made a formal assessment of the design options. It is rare for there to be only one potential design solution and the Court will expect to see detailed consideration of practical alternatives before imposing inconvenience, disruption, and loss of amenity to the neighbouring land.

The access order can stipulate the way the works are to be undertaken and the Court will seek to balance the interests of the parties. Project timelines and detailed analyses will be necessary. For example, for the contractor to erect all scaffolding at start of the project and dismantle at the end, effectively depriving an Owner of their garden for six months for works lasting only three days, would fail a test of 'reasonableness'.

The Court can also award compensation to the neighbour for direct loss, damage and injury, as well as for indirect losses such as privacy and inconvenience. All Owners have a right to protection and compensation payment for direct damage, injury and loss. This should not be confused with a 'premium payment' for financial advantage (e.g., if better access through a neighbour's site reduces the cost of the work) which is a separate concept.

There are important differences between the commercial and residential situations. The Act deals with potential premium payments by taking into account the financial advantage, i.e., lower cost, that the repair works benefit from through better access. Commercial property Owners seeking repair access can therefore be required, by the Court as part of an access order, to make a payment to a neighbour.

The situation changes if the works are being undertaken to 'residential land'. This term is broader than domestic, so a commercial housing landlord, housing developer or other business entity undertaking repair works to residential land is not required to make a premium payment for financial advantage.

When dealing with an application for an access order, the support required from the design team is usually:

- Evidence of compliance with Health and Safety obligations such as CDM 2015[15]
- A technical report confirming that the property is defective and in need of repair (i.e. evidence of the subsidence)
- A design solution that satisfies the test of reasonableness
- A schedule of condition or the ability to prepare such a record once the order is granted
- Evidence that warranties and protections are in place

The contractor will be required to provide the following:

- Detailed programme of works to confirm the number of days/weeks/months of access
- A risk assessment and method statement (RAMS) expressly dealing with the cross boundary and access issues
- Evidence of appropriate insurances

Since 1992 this legislation has triggered only one High Court case, demonstrating that the majority of access requests are granted by neighbours at an early stage, or via resolution at County Court level.

C4 The Party Wall etc. Act 1996

The Party Wall etc. Act 1996[14] came into effect on 1 July 1997 an has been updated since, to include sections on electronic communication and the Crossrail and High Speed 2 (HS2) rail projects. Should subsidence be linked to either of these two major rail projects, specialist legal advice should be obtained.

Serving notice of the proposed work is a mandatory requirement for an Owner to trigger the safeguards of the legislation. However, none of these rights are lost should an adjoining Owner simply give written consent. Importantly, written consent at the commencement does not stop the more formal operation of the legislation at a later stage if needed by the Owners.

In the context of subsidence to terraced properties which can result in adjacent Owners needing repair works to be undertaken, effective communication and amicable neighbourly relations can allow necessary repairs to proceed by written consent alone.

The party wall legislation principally authorises the carrying out of work that would otherwise constitute a trespass, and also confers rights of entry. These rights are narrow and only for the works under the party wall process. Practitioners should avoid seeking to make a project fit the party wall definition when a more formal access licence or access order is the legally correct course. This mis-selection can have serious consequences for all Owners should damage etc. arise. It is relevant to:

- Work on an existing wall shared with another property
- Building on the boundary with a neighbouring property
- Excavating near a neighbouring building

If the work falls within the scope of the Act, the appropriate notice must be served on all those defined by the Act as 'adjoining owners'. Serving notice triggers a legal process enabling an Owner to resolve disagreement with adjoining Owners.

Reaching agreement with any adjoining Owner on party wall matters does not remove the possible need for planning permission, Building Regulations approval or any other statutory requirement.

Some of the aspects likely to relate to subsidence are detailed in the following sections.

C4.1 Scope
The Act applies to:

- Work at or astride the boundary line between properties
- Various work directly affecting an existing party wall including:
    - o Extending it
    - o Underpinning it
    - o Rebuilding it
    - o Repairing it
    - o Reducing its size
- Excavation, or constructing foundations for a new building, within 3m of a neighbouring Owner's building where that work will go deeper than the neighbour's foundations
- Excavation, or constructing foundations for a new building, within 6m of a neighbouring Owner's building where that work will cut a line drawn downwards at 45° from the bottom of the neighbour's foundations

C4.2 Definitions
A wall is a party wall if:

- It stands astride the boundary of land belonging to two (or more) different Owners
- It belongs totally to one Owner but is used by two (or more) Owners to separate their buildings. Where one party has built the wall in the first place, and another has butted their building up against it without constructing their own wall, only the part of the wall that does the separating is considered to be 'party' – sections on either side or above are not considered 'party'.

The Act also uses the expressions:

- 'Party fence wall' — a wall which is not part of a building, that stands astride the boundary line between lands of different Owners and is used to separate those lands. According to a DLUHC booklet[103], this does not include such things as wooden fences
- 'Party structure' — a wider term which includes a floor partition or other structure separating buildings, or parts of buildings approached by separate staircases or entrances

The Act defines an Owner as anyone (other than a mortgagee) with an interest greater than a tenancy from year to year. This means that there may be several 'Owners' of one property.

C4.3 Notices
An Owner intending to carry out work covered by the Act must give notice in writing of the intended works to all the relevant adjoining Owners. This notice must include:

- The Owner's name and address
- The building's address (if different)

- Full details of what is proposed (including plans where appropriate)
- The proposed start date

In the case of excavations, the notice must also state whether the Owner proposes to strengthen or safeguard the foundations of the building or structure belonging to the adjoining Owner, and it must be accompanied by plans.

At least two months' notice is needed for works to an existing party wall; and one month for a planned new wall or for excavation within the specified distances.

Work to an existing party wall
An adjoining Owner cannot stop someone from exercising the rights given to them by the Act, but can influence how and when the work is done. Under the Act, a person who receives a notice about intended work may give consent within 14 days, or give a counter-notice setting out what additional or modified work they would like to be carried out. If an adjoining Owner does not do either of these things, a dispute is deemed to have arisen.

New wall astride the boundary line
If the adjoining Owner gives a notice within 14 days agreeing to the building of a new wall astride the boundary line, the work (as agreed) may go ahead. If an adjoining Owner does not respond, or objects to the proposed new wall, the building Owner may only build the new wall wholly on their own land. They may place footings and foundations under the adjoining Owner's land. Such footings may not be made of reinforced concrete unless the adjoining Owner's express consent is given.

New wall against the boundary line
By agreement or following publication of an award, the Owner may start work, including placing footings and foundations that extend under the adjoining Owner's land, one month after the notice is served. Such footings may not be made of reinforced concrete unless the adjoining Owner's express consent is given.

Excavation near neighbouring buildings
If the adjoining Owner gives a notice within 14 days agreeing to the work, that work (as agreed) may go ahead. If an adjoining Owner does not respond, or objects to the proposed work, a dispute is deemed to have arisen.

Appointment letters and terms of business
Practitioners representing both the Owner and adjoining Owner/s are reminded of the need to ensure that they hold written terms of business that confirm payment terms, insurance and all the standard requirements of being a consultant. Nothing in the appointment provides exemption from normal business liabilities and typical business obligations. Specialist practitioners with appropriate skills and legal understanding are often known as 'Party Wall Surveyors'.

C4.4 Dispute procedure
The Act envisages that, in most cases, there will be agreement about the proposed works between the relevant Owners. This agreement must be in writing. However, where agreement cannot be reached, the dispute resolution procedures of the Act come into play.

The preferred solution is for the Owners to agree on one Surveyor to draw up an 'award'. If they do not agree on one Surveyor, they must each make an appointment. The two Surveyors thus appointed select a third Surveyor who can act as an adjudicator if the first two Surveyors cannot agree. Surveyors appointed under the dispute procedures must act impartially, taking into account the interests of both Owners. This is not a client/agent relationship. Note that an Owner cannot act as their own Surveyor.

C4.5 Award
An 'award' is drawn up by the appointed Surveyor/s. The award is not a financial payment but a statement explaining what work can be carried out, and how and when it is to be done. The award may also specify how the Surveyors' fees will be apportioned — usually the Owner who initiates the works pays. The award is served on all relevant Owners, each of whom is bound by the award unless they appeal within 14 days to the County Court.

C4.6 Cost of work

The general principle of the Act is that the Owner who initiated the work pays for it. However, there are various exceptions, for example:

- Where an adjoining Owner agrees to the construction of a new wall astride the boundary line and makes use of it, the Owner will contribute to the cost
- Where work to a party wall is needed because of defects or lack of repair for which the adjoining Owner may in whole or in part be responsible
- Where an adjoining Owner requests that additional work should be done for their own purposes

The Owner must pay compensation to an adjoining Owner for damage to property caused by the carrying out of work pursuant to the Act, which includes not only the temporary effect of the works but also diminution in value of the property. Compensation may also extend to disturbance and inconvenience.

The Owner may be expected to pay for the strengthening or safeguarding of the neighbouring Owner's foundations where excavation within the specified distances is proposed.

Where the dispute resolution procedure is called on, the award may deal with apportionment of the costs of the work. The dispute procedure may be used specifically to resolve the question of costs.

Appendix D Mining subsidence

D1 Introduction

The effects of past (and likely future) mining and mineral extraction are evident in many parts of the UK. Construction projects are undertaken throughout the country and, in the past when less control was exercised by Local Planning Authorities on development in mining areas, buildings were erected without precaution against the possibility of subsidence from old mining features. The hazards to the performance of structures from past mining activity and mineral working are generally hidden, but their effect on structural performance can be dramatic.

There are many sources of information on historical and contemporary mining in the UK. Many references relate to specific areas of the country, for example the work by Dines that describes in much detail the history of the metalliferous mining in the South West[104].

A more general study however, was undertaken in 1990 by Arup on behalf of the Department of Environment, that reviewed and reported on mining instability[105]. The study was extensive and identified five mineral types with which mining features could be related:

- Metalliferous (non-ferrous) ores
- Rocks such as sandstone and limestone
- Coal and its associated minerals (e.g., fireclay)
- Iron ore outside coal fields
- Evaporites

Each type represents a grouping of minerals in which similarities of geological occurrence and mining methods are found. This study is an invaluable reference for those wishing to initially identify the nature and extent of mining activity in different regions of the country.

The map of Great Britain (Figure D1) was prepared for the study and highlights the principal areas in which the various types of mining are found.

Given the extensive and buried nature of the hazards presented to new and existing buildings, it is essential that a thorough desk study is undertaken in order to identify the probability of structures being affected by mining features. Such a study should incorporate, at an early stage, a Mining Archive Search to identify known and recorded features associated with the mineral extraction.

A desk study and search of available information will include reference to some of the more general publications (e.g. Dines[104]), but should also incorporate a review of specific information. This is usually found in local archives and record offices. Local specialists in mining history and methods may have built up a detailed database of mining activity related to a specific mine or source of extraction. These specialists are often an invaluable source of data and knowledge about the features which are likely to be encountered on a construction project.

Mining archives can provide extensive details of the historical development of an area, including known mining features. Due to the nature of the mineral extraction and the way in which the various mining industries developed, the locations of many older mining features were not recorded. It is these unrecorded workings which pose the greatest hazard to the safe development of a site.

It is necessary therefore, when construction is undertaken in a known mining area to carry out not only an extensive desk study and archive search, but also a site-specific investigation using appropriate techniques. This is required to identify the exact location of unrecorded features and to provide information necessary for the safe design of foundations for the proposed structures.

The following sections give a general overview of two of the principal types of mineral extraction. Coal mining subsidence is discussed first. Secondly, a summary of metalliferous mining subsidence is presented.

Figure D1: Mining areas of Great Britain

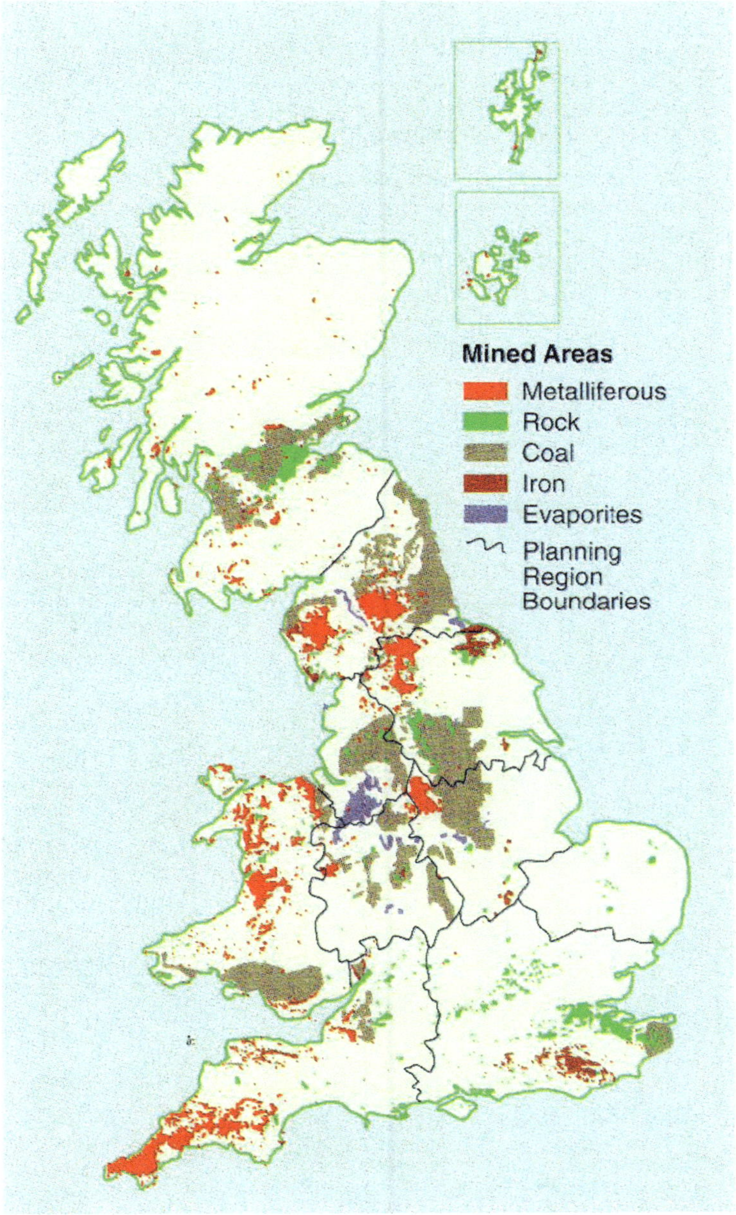

These sections provide an overview of the nature of the problems which can be encountered at sites which have been affected by past mining. There is no substitute for detailed desk study and archive search, backed up by site-specific investigation.

Further guidance is available in CIRIA's *Abandoned mine workings manual*[98]

D2 Coal mining subsidence

D2.1 Introduction

Mining subsidence is the result of the action of gravity on strata rendered unstable by the withdrawal of their natural support over a sufficiently large area. This subsection is a very basic exposition of the technical aspects of subsidence resulting from mineral extraction, particularly coal mining. Considerable information is available in published literature[106] and from mining authorities such as The Coal Authority.

Subsidence, starting at seam level, extends vertically and horizontally with varying time lags to the surface where disturbed ground sinks slowly at different rates until, in the course of time, stability of the whole mass is restored at a lower level. This part of the appendix deals with subsidence due to:

- Modern longwall mining, which is amenable to prediction and analysis
- Pillar and stall (room) mining, which is a more traditional process used prior to the middle of the 19th century, and can affect surface stability in areas of old coal workings. This method is unpredictable in terms of timing and magnitude of ground movement

In the early 1900s, it was known that mining subsidence varied with the size and shape of the underground workings but no generally acceptable theory had been proposed. Since then, considerable progress towards a better understanding has been made by the recognition of basic geometrical relationships between some of the principal factors in mining subsidence, although unsolved problems remain to confuse the prediction of events in specific physical conditions.

Longwall mining procedures can be adopted to minimise surface damage, as the development of surface curvatures and slopes with their associated lateral movements can be predicted with reasonable accuracy. This technique can be used to reduce subsidence damage.

Structural precautions, on the other hand, have not been wholly successful and many problems remain to be tackled, especially in the field of soil-structure interaction, before structures can be adequately designed to accommodate or resist ground movements.

With the accumulation of research into mining areas over the past 60–70 years, empirical rules have been established which relate surface movements to such factors as:

- Seam thickness
- Type of extraction (i.e., mining method and degree of packing, if any)
- Depth of overburden
- Width of working
- Rate of advance

However, surface movements are also governed by less determinate factors which include:

- Character of strata between seam and surface level
- Presence of goaf (the space from which coal has been extracted) in old workings above the new seam
- Presence of geological faults

D2.2 Factors affecting longwall mining subsidence

Mining methods
There are two distinct methods of working: 'longwall', and 'pillar and stall'. In longwall mining, large areas of coal are completely extracted by a number of more or less broad working fronts as the mine face advances. In pillar and stall mining, coal is extracted in 'rooms' and only 30–50% may be removed in the first operation. The walls may be extracted later or left as pillars to support the roof.

The subsidence diagram
In planning and recording their work, mining engineers construct graphs relating surface subsidence to extraction.

Angle of draw
The extent of surface subsidence is defined by the 'angle of draw', measured between the vertical and the line joining the edge of the movement zone on the surface with the edge of the working.

The maximum angle is not developed under all mining conditions in level strata. As confirmed in practice, the smaller the width of face, the smaller the amount of subsidence produced at the surface. By considering faces of progressively reduced width, a stage is eventually reached when a longwall face is too narrow to produce subsidence at the surface (e.g., a single roadway). At this point there is no subsidence and hence no angle of draw.

Critical area of extraction (area of influence)
If from a surface point P angle of draw lines are plotted downwards on all sides, the area subtended in the seam is called the 'critical area of extraction'. As the dimension of the critical area in any particular case depends on the depth of the working and the value of the angle of draw, it will be appreciated that, all other things being equal, the area of influence increases with depth.

The area of extraction affects the subsidence profile considerably. Since point P suffers its maximum possible subsidence when the whole of the critical area is extracted, it is convenient to call this 'full subsidence'. With supercritical areas, where the goaf (extraction) is wider than the critical value, not only does point P suffer full subsidence but other points nearby subside equally, thus forming a flat-bottomed basin. Conversely, when a panel is worked with a width less than the critical value, this subcritical area causes a degree of subsidence less than the maximum possible, and is referred to as 'partial subsidence'.

Seam thickness
The greater the thickness of seam extracted, the greater the subsidence, and it is common practice to express subsidence as a percentage of the seam thickness. Care is required however, to ascertain the true average seam section extracted.

Stowage
Subsidence may be reduced by replacing the extracted coal with waste material. This operation is called 'stowage'. The material is packed tightly in place before appreciable settlement has taken place and may be done by hand or, more effectively, by pneumatic, hydraulic or mechanical means. It is obvious that any material imported into the goaf must, to some extent, reduce the amount of convergence of the roof and floor and thus reduce subsidence at the surface. However, stowing cannot eliminate surface subsidence completely, since in almost all cases the roof subsides 5–10% of the thickness worked, before any packing material can be introduced. The stowing is invariably more compressible than the coal removed, so it is impossible to prevent some degree of surface subsidence over longwall workings using current techniques.

Width/depth ratio
For a given W/D ratio and seam thickness (t) field observations have shown that the maximum subsidence is practically the same irrespective of depth, but ground strains are affected by amplitude of subsidence and depth.

Geology of the overburden
In order to predict the effect of geological characteristics, an intimate knowledge of the strata comprising the overburden is essential. McTrusty[107] observed that strong massive beds of sandstone or limestone may subside erratically in large blocks between joints, thereby leading to irregular subsidence at the surface, whereas shales and clays tend to yield more uniformly. A thick bed of sand or sandy clay may, by internal adjustment of its particles, absorb erratic displacements in underlying beds and thus reduce irregularities at the surface.

Old coal workings
The effect on subsidence caused by the presence of goaf above the seam being worked is difficult to assess, especially where pillars, ribs and barriers have been left in old workings. Movement of the old workings is superimposed on the normal subsidence and Whetton[108] considers that the major problems concern prediction of the magnitude of the strain on the surface, both in position and direction.

Presence of faults
The most erratic surface changes occur in the presence of a geological fault. This natural plane of shear modifies the normal effects of draw and may cause abrupt changes of surface level with spectacular effect. Faults are also susceptible to reactivation during mining subsidence[109].

D2.3 Estimation of subsidence from longwall mining

The purpose of this section is to describe the principles of subsidence due to longwall mining, since it is essential that the Engineer who sets out to design a structure in a mining area should be familiar with the basic concepts.

Subsidence factor
When all the effects discussed in the previous subsection have been taken into account, it is usual to express the estimated maximum subsidence as a fraction of the seam thickness extracted. This fraction is known as the

'subsidence factor'. An estimate of subsidence factor can be made for longwall workings with strip packing – provided the seam is level, the width of panel is constant and the face has advanced through the critical area for the point considered.

While the greatest subsidence factor is 0.84, the value may be higher for full caving but is certainly lower for solid stowing. Thus, the method is strictly limited in its scope and, even when applied to strip packing, there is no doubt that the factor is subject to variations according to local conditions.

Strains and displacements
When a trough of subsidence is formed, the centre part subsides vertically only, and the remainder moves inwards in addition to moving downwards. Observations show that the horizontal components of the movement at individual surface points in the subsiding area are not equal in amount, and the differential displacements cause tensile and compressive strains according to which of the inner or outer points is displaced to a greater extent. The tensile strains occur mainly around the edges of a subsiding basin, and the compressive strains more or less directly over the worked-out area. The horizontal displacements reach a maximum at points approximately vertically over the edge of the worked-out area and are zero at the extremities and at the centre of the subsiding basin. One phenomenon, which is a direct outcome of tensile strain, is the occurrence of breaks or fissures in the ground surface, usually at points of maximum tensile strain. These fissures tend to appear where a stratum of hard rock is present, at or near the surface, but may occur also in softer strata if the tensile strain is great.

To obtain full subsidence, the width of working must be equal to the diameter of the critical area but it has been observed that the maximum strains are reached when the width is approximately equal to the radius of the critical area. As the width of working increases, the tensile strain remains constant at its maximum intensity, while the compressive strain reduces in value and gradually spreads out into two zones when the full critical area has been worked. Under these conditions, it is commonly observed that the compressive strain tends to reduce to the value of the tensile strain, in accordance with the long-established principle that the amount of tensile strain in two zones must more or less balance the compressive strain over the goaf, whether in one or two zones. For a critical area of extraction there exists a point which is strain-free and when a supercritical width is worked, this strain free point becomes a zone.

Travelling or dynamic movements
Development of subsidence is a three-dimensional process, and as the face advances it initiates a wave of subsidence at the surface where tensile and compressive strains are experienced alternately.

Beevers and Wardell[110] found that the intensity of the travelling strain bears a direct relationship to the amount of subsidence and so, for a face equal in width to the diameter of the critical area, the maximum travelling strain develops over the centre of the face, where the greatest subsidence is also developed. To right and left of the centre, both the subsidence and intensity of the travelling strains decrease.

Travelling strains are greatest when the width of the advancing face equals or exceeds the diameter of the critical area. Therefore, a surface structure placed in the central zone of a supercritical area of extraction would be subjected to maximum travelling strains, although eventually it would return to a strain-free condition at a lower level than originally located.

Slope
An estimate of surface slope (or 'tilt') is required to determine the size of structural units and the need for expansion gaps to minimise the risk of damage. Mining engineers calculate slope between adjacent subsidence stations, 15–30m apart, by dividing the differential subsidence by the horizontal distance between stations.

The time factor
From the point of view of surface damage, the duration of ground movement is a factor of concern to mining Engineers and Surveyors who have to deal with damage claims and repairs. It is important to know when a particular structure is likely to be initially affected, when the danger will be most acute, and when the movement will be virtually complete. Workings must be properly phased to avoid cumulative strains. A knowledge of the relationship between time and subsidence is therefore essential.

Wardell[111] clarified the situation by introducing the 'subsidence development curve'. This curve indicates the subsidence of a given point on the surface for successive positions of a face passing through the critical area of the point. Wardell represented the time factor by the rate of advance expressed in terms of the radius of the critical area and, from eleven case histories, concluded that the development of subsidence depended primarily on the position and rate of advance of the face within the critical area.

D2.4 Surface damage from longwall mining

General aspects

The effect of subsidence on structures and services are dealt with elsewhere in this guidance. Minor effects include cracking of plasterwork and slight jamming of doors and windows. More serious effects are fractures in brickwork and distortion of steel or concrete frameworks which may lead to structural failure.

Roads, railways and waterways have subsided as much as 6m where a number of seams overlying each other have been worked. This has necessitated the building-up of banks of waterways, followed by the raising of bridges to maintain headroom. Areas of low-lying land become inundated with water to form flashes and meres, thereby detracting from land values and rendering existing buildings untenable. Backfalls may be created in pipelines due to variations in vertical subsidence.

In zones of compression, the ground has been observed to form ridges some 1m wide and 200mm high, while in tension zones cracks of width exceeding 150mm have been measured. As a result of these horizontal ground displacements, sewers and other pipelines may be buckled and fractured, and basements and manholes distorted and crushed.

With regard to surface structures, it is well established that the tensile and compressive ground strains are the main cause of damage. This is confirmed by the fact that the greatest damage does not necessarily take place where maximum subsidence occurs but rather where the curvature of the subsidence basin is greatest. The degree of damage suffered by the structure however, is not simply dependent on the intensity of compressive or tensile ground strain, but also on the dimensions and orientation of the affected structure with respect to the advancing face, the subgrade forces at the foundation interface and the flexibility of the superstructure. If the ground is being compressed at a uniform strain of, for example, 1mm/m, the damage resulting from this compressive strain may be slight in the case of a small building, but if the whole of the shortening takes place at a single weak point in the structure, a long building, such as a terrace of houses, may suffer severe damage.

In addition to horizontal strains, tilt can also adversely affect buildings, especially tall structures such as chimneys. Here the structure may lean to such an extent that it becomes unstable. This occurrence however, is rare and nearly always due to several seams being worked on one side of the structure only. The remedy is to work coal on the other side, although this is not always convenient.

Slight or moderate tilting, unaccompanied by appreciable strains, is often tolerable or even unnoticeable, but certain industrial processes can be upset by slight departures from the horizontal, and provision for jacking is sometimes made where the possibility of levelling by underground extraction is impracticable.

In general, although damage due to mining can be very severe, it appears that aesthetic damage is a prime concern. Although structurally sound, a property which leans or exhibits plaster cracks can cause serious distress to its occupants. For this reason a knowledge of ground strains and settlements present at the first appearance of cracks is of particular importance.

Although surface subsidence due to mining is three-dimensional in character, in order to study the threshold or limiting movements in relation to degree of damage, it is convenient to consider the vertical and horizontal components separately, in spite of the fact that they occur simultaneously.

Vertical settlements

Curvature of the ground surface usually gives rise to differential settlement between parts of a structure and, unless the structure is statically determinate or perfectly flexible, the distribution of the supporting forces will be changed. This will, in turn, produce stress changes within the structure and these may ultimately result in structural damage. In ground settlement problems, the Engineer has two principal tasks:

1. To evaluate the allowable differential settlement the structure can withstand without experiencing damage.
2. To predict what total and differential settlements can be expected at a given site.

Field observations show that the magnitude of differential settlement which will cause damage to a building cannot be predicted from a theoretical computation, since the behaviour of the building is influenced by a number of factors which are difficult or impossible to take into account in theoretical calculations — such as interaction between frame and cladding, the varying rates of subsidence and the redistribution of the loads. As such, allowable settlements have to be decided on the basis of experience derived from observations on similar types of structures.

The second task for the Engineer concerns the prediction of differential settlement at a given site. Settlements occurring as a result of mining operations are, of course, additional to settlements existing before mining operations commence.

Horizontal displacements
Ground strains at the surface arising from differential horizontal displacements are not transferred directly to the structure. The ground movements impose a force on the structure which, in turn, causes structural strain. The fact that ground strains do not reproduce identical strains on the structure sited on or in the ground was not generally appreciated by planning and mining Engineers until the late 1950s.

At present, the only guide to estimating damage is based on many years of observing countless buildings and, despite the fact that several factors such as soil type, subgrade force and structural flexibility have not been included in the observations or in the empirical analyses, simple relationships between ground strain and damage have proved to be of value. When new extractions are planned, a reasonable assumption of the damage likely to be caused to overlying surface properties can be made. Any particular precautions indicated can then be taken, whether this be in the form of modifications to the structure or to the proposed mine workings.

Most buildings can withstand a certain degree of tensile or compressive ground strain without apparent signs of damage, and observations indicate that cracking of internal plaster and jamming of doors and windows tend to occur when tensile and compressive ground strains exceed 0.4–0.8mm/m, respectively. These figures lead to the commonly held view that most buildings appear to be about twice as strong in compression as in tension. Compressive ground strains up to 6mm/m have been recorded and these would, without doubt, cause severe surface damage. Orchard[112] compared several cases where the ground strains have been measured precisely and the structural damage has been photographed and classified according to an accepted scale of damage.

D2.5 Ground movement due to old coal workings

General
Although ground movements due to longwall mining may be predicted with reasonable accuracy from the empirical relationship derived by British Coal, those due to old coal workings can be assessed only on the basis of simple models of behaviour[113], records of collapse and from experience of building performance within districts underlain by former workings. This subsection has been included to provide background information for the Engineer confronted with the problems of building development within such districts, but it must be emphasised that specialist geotechnical advice should be sought at the preliminary stage of the site study.

Thousands of old mine workings have been charted in the UK, but unrecorded workings exist (as legislation making it compulsory to keep plans of old workings and shafts was not introduced until 1872). These workings are largely the product of coal mining operations in the 17–19th centuries when coal pillars were often left to support the overlying strata. Note, that old workings due to the mining of silver, lead, shale and salt also exist, and that iron minerals and refractory clays were extensively mined at the outset of the Industrial Revolution. However, the following information concentrates on the major problems related to coal mining.

At shallow depths, coal pillars may continue to support the overlying strata for a considerable time, perhaps several hundred years, before gradual deterioration leads to collapse of roof strata between pillars, leading to the upward migration of cavities. Pillar collapse may also occur in certain circumstances. Where cavities reach the surface, local subsidence occurs, often quite suddenly.

Access and air shafts associated with former mineral workings have often been partially and loosely infilled and covered such that their exact location is extremely difficult to establish. In some cases the mine shafts were completely infilled and/or sealed at the surface of the bedrock by means of a reinforced concrete slab. In many cases however, a timber platform was erected at the surface of the bedrock and heterogeneous fill materials deposited between the platform and ground surface. With the passage of time the timbers decay, and sudden failure can be experienced in the form of local and rapid drawdown of surface materials.

The disturbance experienced at ground surface by a structure due to collapse of shallow workings or old mine shafts is often related to the nature and condition of the superficial deposits. Loose water-bearing essentially non-cohesive deposits such as sands and silts may be expected to accentuate the ground disturbance caused by shallow mineral workings, in direct contrast to the relatively better ground support conditions provided by a stiff cohesive deposit such as glacial till.

Serious structural failures are possible if buildings are founded above hidden unstable shafts or collapsing pillar and stall workings[114, 115]. Despite there being little information on the subject, the coal seams in question are extensive, often underlying major urban areas created as a result of the Industrial Revolution[116] (Figure D2). These areas continue to expand and simultaneously undergo central redevelopment, which may involve the construction of buildings founded over strata partially removed by mining.

Figure D.2: The coalfields of Britain in relation to the main centres of population

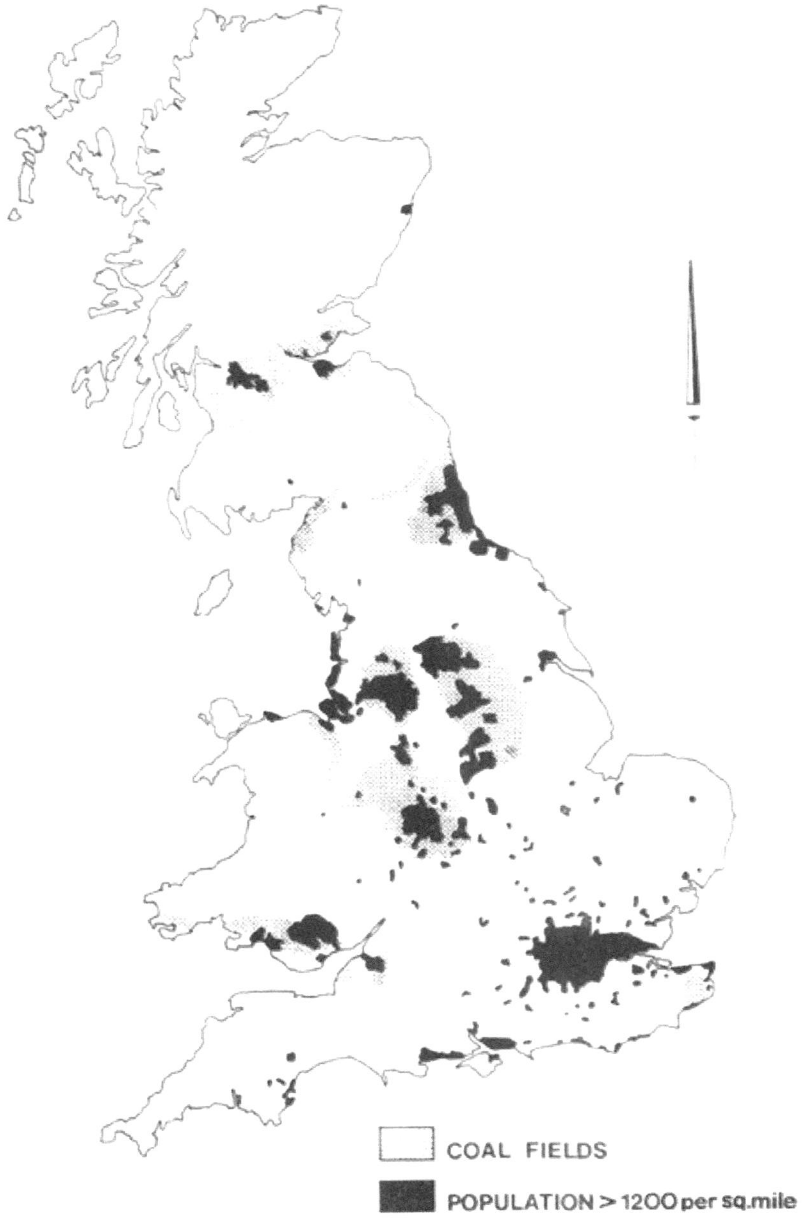

As workings age, they become more susceptible to collapse due, for example, to the slow but steady erosion of material in the workings by groundwater. Operations on the ground surface such as piling can also affect the stability of the workings. Construction over mined areas should be considered very carefully and particular emphasis placed on site appraisal and ground investigation[117].

Reference should be made to the British Geological Survey for information concerning the geology of the district within which it is proposed to erect any structure[43]. Reference should also be made to The Coal Authority who possess the archive of coal mining abandonment plans[97]. Difficulties can be experienced in obtaining detailed information on the extraction of other types of minerals, particularly in the case of limestones, ironstone and metallic-type ores. Some information is held by The Coal Authority and local mining groups.

The Coal Authority are a statutory consultee for planning applications in high-risk development areas. Planning conditions may be applied with respect to the requirement for investigation and treatment of old coal mine workings.

Surface stability of old workings
Old coal workings are prone to collapse, which can occur sporadically on a site over a period of time, and at any time after mining is abandoned, unless the workings are stabilised prior to development. The collapse of old workings is not necessarily dependent on surface load being applied by foundation construction (Figure D3), although such engineering activity may have a marked detrimental effect on the conditions leading to instability.

Figure D3: Mine shaft collapse

Uncollapsed or partially collapsed workings, if dated before 1872, are not likely to be recorded. Even where records exist there is no guarantee of accuracy in the description of pillar and stall geometry and neither can assumptions be made of the deterioration of the workings.

At present it is impossible to predict the timing of any collapse, but a careful site investigation to assess the degree of collapse or roof convergence to date, combined with a knowledge of the mining method and general geometry of the workings, do permit an assessment of the maximum subsidence likely if total collapse occurs. The magnitude and distribution of ground strains however, cannot be forecast with any accuracy and these can have the greatest influence on the subsequent performance of a surface structure.

D2.6 Ground movement due to mine entries

Introduction
Knowledge of the shape, plan dimensions and depth of shafts and 'adits' (horizontal or near horizontal passages) is important in gauging the consequences if collapse occurs, or in establishing precautionary measures, e.g., backfilling, capping or grouting, etc. that may be required to prevent subsidence. Shaft shapes ranged from circular and oval to square and rectangular, and while circular was the most popular in the UK, certain regions preferred square and rectangular sections. It is also useful to be familiar with the types of mine shafts, since knowing the function of the shaft may make it possible to infer its proximity to other shafts or the workings. Mining adits may be treated as shallow workings to predict their subsidence effect and any treatment.

Mine shaft fill
Old shafts will be encountered in three conditions of abandonment, i.e., completely filled, partially filled, or open. Many shafts have been only partly filled in, the fill having been placed on a staging, usually timber, within the shaft. The level for the staging varies in practice, but it is often found just below ground level or in the vicinity of a rock head. On occasions the upper shaft lining was removed so that the staging could rest on a secure rim. Late 19th century shafts had the timber replaced by an open grid of iron joists. Experience indicates that large objects such as mine cars or trees were thrown into many shafts to form an obstruction on which to start filling, and sometimes large objects have accidentally bridged the shaft, creating an obstruction for the fill.

Note, that historically, there was no standard practice of closing off galleries and drifts, so fill could migrate from the shaft bottom. Also there is no evidence to suggest that fill was deliberately compacted when being placed. Any tendency to consolidate would be due to self-weight, leading to surface depressions. This type of movement can occur many years after filling was completed and can often be hidden by the surfacing above the shaft, sometimes leading to voiding between the fill and the surface above the shaft. This can lead to rapid failure of the surface.

Capping
Shafts that were not filled have traditionally been made secure by fencing or capping. In practice many methods of capping have been noted, including the use of abandoned haystack boilers from Newcomen engines, and brick cupolas. Until the nationalisation of the coal industry, there were few legal requirements concerning the protection of shafts.

Shafts that are recorded as being capped to a specified design can still be subject to collapse. Many older cap designs do not conform to modern standards and it should not be assumed that the presence of a cap means that a mine shaft is stable.

Surface stability of old shafts
Experience indicates that loss of support from a large area of the surface may occur suddenly when abandoned shafts collapse. Where a rock head is near the surface and the collapse arises from disintegration of the staging supporting the fill, the subsidence will be severe but localised. In the case of a breakdown of the shaft lining, or where a thick deposit of unconsolidated material overlies the shaft, a large crater may be formed as the deposits flow into the shaft.

Clearly, the possibility exists that disused shafts were not efficiently filled when abandoned. Even if their position and construction technique were faithfully recorded, the process of deterioration of the shaft fill, lining and staging can seldom be quantified. At present, no predictivity capacity exists in relation to the time, magnitude or extent of surface subsidence arising from shaft collapse.

D2.7 Reporting subsidence to The Coal Authority
The Coal Authority may be responsible for subsidence to a property caused by historic coal mining. Under the Coal Mining Subsidence Act 1991[118], a damage notice can be submitted to the Authority. If liability is accepted then repairs and/or remediation of the workings and property would be undertaken by them.

D3 Metalliferous mining subsidence

D3.1 General
This section presents a brief overview of metalliferous mining and its effect on building structures. It is based wholly on experience in the South West of England, although the principles may be applied to this type of mining elsewhere.

The information presented was obtained from practitioners in the region with experience gained from many years investigating subsidence-related problems due to old metalliferous mining features. In addition some data has been obtained from published references on the historical mining of the area.

Many older buildings in mining areas have experienced problems due to subsidence. This can be related to both recorded and unrecorded workings. It is not uncommon to find old structures, which have been constructed directly over backfilled shafts, or other surface and near-surface features, which are actually recorded and noted on old plans of the mines at the time they were abandoned. Some features are also noted on Ordnance Survey maps of mining areas. More often however, mining subsidence is due to unrecorded near-surface or surface workings which have been previously backfilled.

Over time, buildings begin to show signs of damage due to subsidence. Depending on the extent of the mining feature, the damage can be superficial or progressive. Either way remedial works will often be needed, requiring knowledge of the ground conditions and information about the geometry of the mining feature/s, and this will require some appreciation of the types of feature which are associated with the particular type of mining. The following subsection presents a discussion of typical surface mining related features, together with a simple schematic drawing of the various forms of mining structure which can be encountered.

D3.2 Typical surface mining related features

Introduction

This subsection describes the typical early workings related to surface extraction of minerals, which lie above the deep and shallow adit elevations of metalliferous mines. Such surface features are rarely recorded and are responsible for many of the subsidence problems that are encountered in the South West of England and other parts of the country.

A simple schematic (Figure D4) shows the typical features of early workings related to the surface, and lying above the deep and shallow adit elevations. It is these workings which are very rarely recorded.

Figure D4: Schematic view of mine workings (Cornwall Mining Services Ltd)

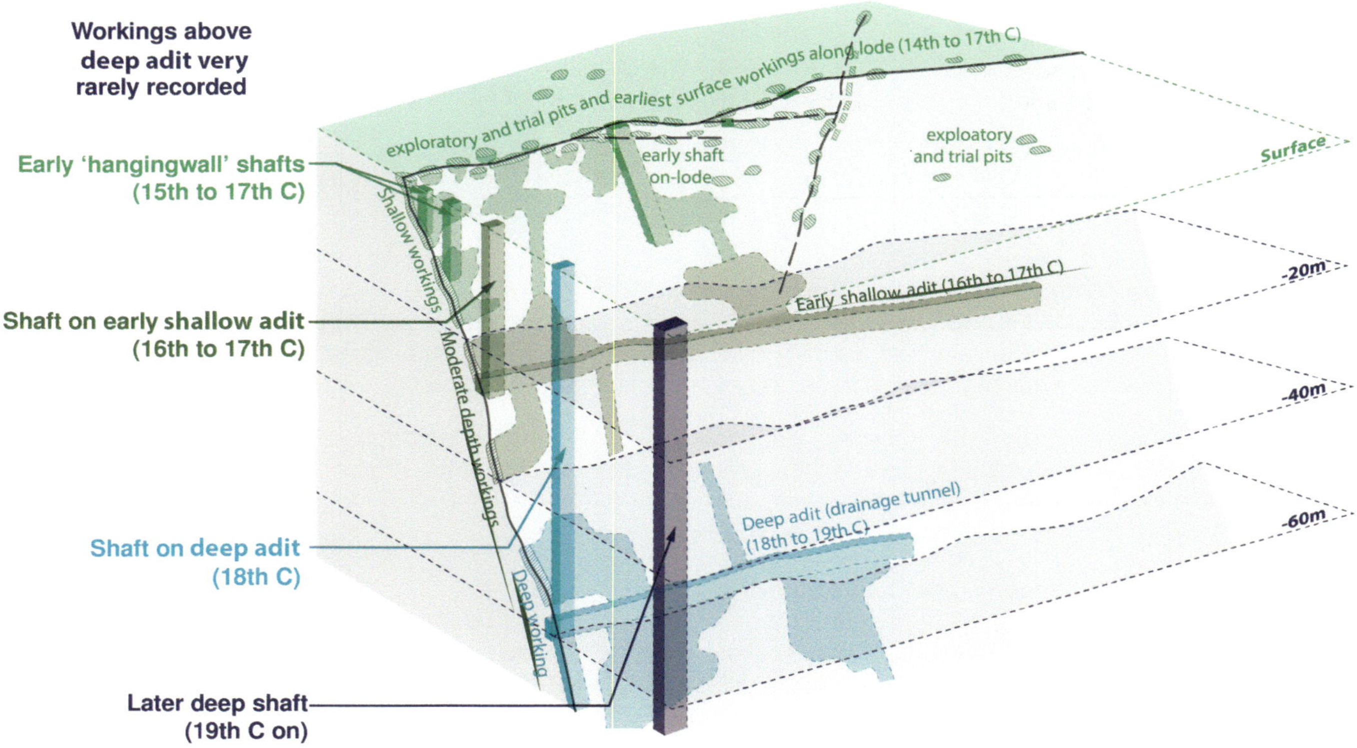

The following subsections describe the various features which may be encountered. The terms described relate to mining in the South West of England. Note, that terminology will probably vary depending on the location of the metalliferous mining, with local terminology having developed with the mining itself.

Early exploratory pits, trenches and tunnels

A very common feature in any mineralised area, there are often hundreds of early exploratory pits, also known as 'costeening' pits, within 1 acre of ground. They are mostly irregular in shape and variable in size, but typically roughly oval in section, and 1.5m by 0.7m to 2.5m by 1.5m, tapering with depth to a base generally within the structured rock head, and typically being 2–4m deep. Very short tunnels ('stubs') can be present. Backfill may typically be of earth and lay, with mixed near-surface rock. The tunnels usually remain open. These pits are often found in clusters, or in rough bands following lode veins or outcrops.

Occasionally, surface trenches have been cut across the trend of the lode (vein of metal ore), or more often along an outcrop by way of testing. The latter may be very variable in width and length, are of similar depth to the exploratory pits, and are effectively the first phase of excavating the lodes. However, they are typically of an ovoid section, and spaced along the lode.

Exploratory near-surface tunnels, driven from deep pits or very shallow 'shafts' can be present, which often follow along or adjacent to a cross-course fault or to a lode vein, and with their base entering the structured but broken rock head horizon, resulting in their roof often lying 1.5–2.5m below the surface. These (mostly still partly open) tunnels are typically 0.6m wide by 1.6m high.

Early lode outcrop workings
If a discovered lode was proved by the previous exploratory methods to be worth following in depth, before the use of shafts, the pits or trenches were simply deepened into often very irregular 'warren like' workings, resulting in very deep pits and slots running into underground workings, often inter-connecting between the surface workings. If a lode was very rich at surface, a near-continuous deep slot or trench would result, with no intermittent pillars remaining. Where the lode was wide, or a close group of veins was present, these workings took for the form of a surface quarry type lode working. These workings were mostly backfilled when moorland reclamation was carried out, using the adjacent waste tips of earth, clay and mine waste, together with imported materials.

Shafts
There are various types of shaft, related not so much to their function, but to the historical or technical period of working, and are discussed in a chronological sequence in the following three subsections.

Early exploratory or trial shafts
These features generally lie in areas where either a lode had been discovered that was being followed, or traced after being lost, or where exploratory pitting or 'costeening' had indicated lode veins. They may in their simplest form be an extension of any early pit, and may be very irregular in their shape, often tapering from a collar 1.5–3m in diameter, down to a roughly square or rectangular section, typically 1.3m by 0.7m. Circular shafts are rare in the South West. Backfill material within these shafts is predominantly of earth material, with near-surface clay and rock. Stubs may be present very near to surface.

Early lode shafts
Mostly following the earliest lode outcrop workings, but sometimes concurrent with them, are small- to medium-sized, roughly square or rectangular shafts, typically 1.5m by 1.5m, and 1.2m by 2m in cross-section, with nominal surface 'coning'. These lie immediately along or adjacent to the lode outcrop, unless the lode is of moderate to shallow dip, in which case they lie close to the 'hanging wall' side of the outcrop. Backfill material within these shafts is often of clean or earthy cobbles, originating from depth, but may also be of earthy, near-surface rock and clay.

Early adit shafts
In the construction of shallow drainage tunnels between the lode workings and an adjacent valley side or bottom, vertical shafts were sunk along the course of the tunnel to facilitate both rock extraction and ventilation. They may typically be of rectangular section, 0.8m by 1.5m, with no significant 'coning', and may still be partly open, being often originally 'sollared' (platformed) by timber mostly near to surface. The early shallow adit shafts may typically be spaced at around 40m apart, the spacing increasing (often doubling or more) with increased depth. Backfill may be very variable, as these shafts have often been filled following later collapse of the timber sollar.

With the advent of steam power, mining rapidly progressed to far deeper levels than had hitherto been possible, with the need for new, deep shafts. For vertical and near-vertical lodes, these may be continuations of earlier shafts 'on-lode', but for less steep to shallow dipping lodes, new shafts were sunk well into the 'hanging wall' side of the lode outcrop, being vertical to the lode intersection, and then down the lode 'on-dip'.

These mostly rectangular shafts vary in size, depending on their period and type, and may typically be between 2.4m by 1.3m, and 3.5m by 1.8m. Backfill can be very variable, and may be of household or industrial waste, as well as mine waste. These shafts can still be open, and may occasionally be sollared near to the surface, either with timber or occasionally with arched stonework. More recently, many have been secured by mass concrete plugging, reinforced concrete capping, or protected by steel grids or 'copes'.

Shaft and engine house-related features
Various pits, slots and sumps can be present within engine houses, or adjacent or near to engine houses and shafts, and are not easily described as typical features. They may occasionally be deep, and cut into the rock, and may often be stone walled against the cut sides.

Non-mining related quarries and pits
In addition to recorded pits and quarries, unrecorded pits and quarries can be present, which may be up to several metres in depth. These can sometimes be backfilled and reclaimed.

Early tin stream works
Any part of a valley alluvium or other tin-bearing superficial deposit may have been worked in the past, and the general effects are usually of a widespread disturbance of the natural sequence of the deposit. Deeper, subsurface gravel workings are not indicated in the study area, although shallow exploratory tunnels can be present.

Surface waste tips
These may appear to be obvious, but can be present with no noticeable remaining surface signs. The materials can vary from coarse rock to fine processed sands, and often contain significant metallic ore contents. They have often been partly or wholly removed.

Shallow adits (drainage) or exploratory tunnels
In addition to the near-surface tunnels as previously described, these (mostly still partly open) near-horizontal tunnels run from the valley side or bottom, and for the first section may be very close to the surface, depending on the valley side slope, or whether the tunnel passes along the valley bottom before running into the slope. The tunnel may typically be 0.7m wide by 1.7m high. In the lowest section, an adit may well be in the form of a timbered or stone-lined and roofed cut trench, or open trench or ditch.

D3.3 Investigation of metalliferous mining subsidence
In the introduction to this appendix, reference was made to the requirement for a desk study to identify the extent of the known workings in an area. The desk study and Mining Archive Search are the first step in assembling sufficient information to enable a more detailed site investigation to be undertaken. The desk study should seek to identify as far as possible the nature and extent of any known and recorded features which might affect the building in question.

From geological information and data, and available mining records, sections through the site can be drawn to identify the likely locations of lode outcrops. This provides a first step towards locating the possible worked or backfilled areas which might have influenced the structural behaviour of buildings in mining areas.

A detailed structural inspection carried out in accordance with the methods described in Chapters 6 and 7 will also provide detailed evidence about the nature of structural movement and the likely mechanisms which have affected the construction. The combination of structural inspection and desk study may provide sufficient data to be able to interpret the subsidence. Additional site-specific investigation will be needed however, to provide conclusive proof of the nature of the features which caused the subsidence.

Once information is available from the desk study, the Mining Archive Search and the structural inspection, a more detailed site investigation may be planned. This may include trial holes, trial trenches, probe drilling or borehole drilling.

Hand probing tools have been used in the past for investigative purposes, but caution should be exercised when using such methods as they can lead to erroneous results, particularly in ground containing cobbles and boulders. The use of hand probes is not recommended.

A very useful development in current drilling techniques is 'measurement while drilling' (MWD). If this is used with borehole or air flush probe drilling, a continuous read-out is produced of all the drilling parameters. These include rate of bit rotation, penetration rate, bit pressure, torque, etc. The advantage of such techniques, particularly when using open flush holes is that they can provide high quality data relating to areas in which backfilled voids are present.

Appendix E Building Regulations

This is a brief overview of the Building Regulations as they might apply in England and Wales, Scotland and Northern Ireland where relevant to subsidence. It is not intended to be full coverage and advice should be sought from your building control body for individual circumstances.

E1 England and Wales

The Building Act 1984[119] gives the Secretary of State and Welsh Ministers power to make regulations, requires Local Authorities to enforce provisions of the Act, and defines the scope of Approved Inspectors (companies that provide building control as an alternative to the Local Authority). Note, that from April 2024, the term 'Approved Inspector' is largely superseded by 'Registered Building Control Approver' (RBCA); therefore any reference to 'Approved Inspector' should be read as a reference to a Registered Building Control Approver.

The Building Regulations 2010[16] apply in England and Wales to 'building work', but do not apply in Scotland or Northern Ireland where there are different arrangements. Since 2011, separate regulations have been made by the Welsh Government; however, these regulations are similar to those in England and the two are discussed alongside one another here with any relevant differences noted.

The Building Act and regulations are subject to constant review and are updated periodically. It is therefore advisable to check on the application of legislation or guidance to the proposed work.

Developers have a choice whether to use the Local Authority or Approved Inspector for building regulation purposes.

An Approved Inspector is an individual or organisation authorised to carry out building control work; however, they cannot carry out enforcement and must 'revert' this to a Local Authority if agreement cannot be reached with the person undertaking the work.

Local Authorities have a duty to implement the Building Act and the Building Regulations. This means they not only have the power to control building works in their area, but they have an obligation in law to exercise that control, including where an application is reverted from an Approved Inspector.

Building Regulation approvals are not to be confused with planning consents, these are separate and controlled under the Town and Country Planning Act 1990[65].

In some instances, work may be exempt under Building Regulations but not under the Town and Country Planning Act; similarly, work exempt from planning requirements may need Building Regulation approval.

Building work as defined by the regulations includes:

- Erection, extension or material alteration of a building
- Provision, extension or alteration of a controlled service or fitting in connection with a building
- Material change of use of a building
- Work involving the underpinning of a building

amongst other work relating to energy performance and installation of insulation.

Building work must be carried out in a workmanlike manner with adequate and 'proper' materials. Proper materials include materials which are assessed in accordance with the Construction Products Regulation (CPR)[120], have independent certification, or are in accordance with British, European or International standards. Any evidence used should be thoroughly checked to ensure it is applicable to the end use of the product. It is recommended to consult your Local Authority or Approved Inspector regarding the application of any method used to demonstrate compliance.

E1.1 Requirements

The current Building Regulations[16] came into force in 2010 and have been altered numerous times. Since 2011, Building Regulations have been a devolved matter for the Welsh Government. It is therefore important to read the regulations alongside the relevant amendments for England or Wales only, depending on the location of the work. Consult your building control body for further advice on the Regulations as they may apply to the works.

Regulation 3 'Meaning of Building Work' includes the erection or extension of a building, which could include foundations, as well as 'work involving the underpinning of a building'.

Regulation 4 states that building work shall be carried out so that it complies with the relevant requirements of Schedule 1 of the regulations.

E1.1.1 Schedule 1, Part A: Structure

Requirements A1 and A2 stipulate that a building is to be constructed so that the combination of dead, imposed and wind load are sustained and transmitted to the ground. This should be achieved safely and without causing deflection or deformation of the building (or part of the building), or movement of the ground (caused by swelling, shrinkage, freezing, landslip or subsidence) which will affect the stability of any building.

Requirement A3 states that buildings must be built so that in the event of an accident, the building would not suffer collapse to an extent disproportionate to the cause.

The Secretary of State and Welsh Ministers have also produced a series of 'Approved Documents' which provide practical guidance on demonstrating compliance. If the guidance contained within these documents is followed this will be evidence generally demonstrating compliance with the regulations. However, they will not automatically guarantee compliance in every situation and solutions should consider the wording of each requirement, especially in relation to less common building situations.

Approved Document A[121, 122] relates to building structure. The first parts of this document, give practical guidance on matters relating to loading and ground movement.

At the beginning of Section 2 there are a series of definitions including:

- Dead loads (permanent actions), which are the load of the building itself
- Imposed loads (variable actions), which are the loads that can be brought into or put onto the building such as those due to snow, wind, furniture, machinery or the people who use the building

These are the distinct types of loading for which a building or part of a building must be designed to support or resist. Guidance on construction of certain residential buildings or other 'small buildings' follows. Section 2E of Approval Document A relates to strip foundations of plain concrete and contains guidance on strip width to linear load. There are no other specific requirements relating to foundations or underpinning in Approved Document A.

Section 1 of Approved Document A lists codes, standards and references for structural design and construction of all building types that can be used to demonstrate compliance with Requirements A1 and A2.

Reference is also made to a series of Government-sponsored documents on the matter of subsidence. It should be noted that the standards referenced in Section 1 of the document are not exhaustive and other technical information, test data, codes etc., can be used to prove any proposal or form of construction; provided that the building control body agree its relevance to the case in hand.

E1.1.2 Other requirements

The other Approved Documents in England[123] and Wales[124] are:

- Approved Document B: Fire safety
- Approved Document C: Site preparation and resistance to contaminants and moisture
- Approved Document D: Toxic substances
- Approved Document E: Resistance to the passage of sound
- Approved Document F: Ventilation
- Approved Document G: Sanitation, hot water safety and water efficiency

- Approved Document H: Drainage and waste disposal
- Approved Document J: Combustion appliances and fuel storage systems
- Approved Document K: Protection from falling, collision and impact
- Approved Document L: Conservation of fuel and power
- Approved Document M: Access to and use of buildings
- Approved Document N: Glazing — safety in relation to impact, opening and cleaning (Wales only)
- Approved Document O: Overheating
- Approved Document P: Electrical safety — Dwellings
- Approved Document Q: Security
- Approved Document R: Physical infrastructure for high-speed electronic communications networks
- Approved Document S: Infrastructure for the charging of electric vehicles (England only)

Approved Document C may be particularly relevant for treatment of ground contamination and reconstruction of foundations, walls and floors to resist moisture ingress. Approved Document L will also need to be consulted for guidance on thermal insulation of ground floors and walls.

E1.1.3 Regulation 7(2)–7(4): 'ban on combustible materials'

Although currently outside the scope of this guidance, combustible materials are banned in both England and Wales in the external walls of certain buildings with a storey height of at least 18m above ground level, or some lower buildings that may be attached. Specialist advice should be sought.

E1.2 Exemption

Certain exemptions apply to some building types or certain types of small extension. Additionally, certain buildings of statutory undertakers or other specified bodies are exempt, as well as buildings belonging to the Crown. Your building control body can give you further guidance on whether a building is likely to be exempt.

E1.3 Application

A choice of either a Local Authority or an Approved Inspector can be made for control of 'building work'.

In the case of using a Local Authority, submissions may be through the building control approval with full plans or 'building notice' route. For both routes, regular inspections in accordance with a site service plan are undertaken to review progress of works and establish whether compliance can be demonstrated; ultimately leading to a 'completion certificate' if satisfactory. For building control approval with full plans, the authority will additionally undertake a review of submitted details and issue an approval (with possible conditions) if they are satisfied they show compliance. Limited details are given with a building notice, and no formal review or associated approval is given.

Where an Approved Inspector is being used, the person carrying out the work and the Approved Inspector will jointly give the Local Authority a notice advising that the works will be overseen by the Approved Inspector; this is known as an 'initial notice'. The Approved Inspector will undertake inspections throughout the works and issue a final certificate if they are satisfied that compliance is demonstrated on completion. An Approved Inspector has no enforcement powers; therefore, if they are unable to resolve non-compliance the application must be transferred to the Local Authority for enforcement action, this is known as 'reversion'.

It is important to discuss and agree the submission method with the building control body being used, because some methods may have rules or conditions that prohibit or restrict their use.

E1.4 Regularisation

If works have been undertaken without a valid Building Regulation submission, the Building Regulations provide for a Local Authority to regularise the work. They can issue a certificate where they are satisfied (as far as they can determine) that relevant requirements have been fulfilled. To determine compliance, opening-up of the building fabric may be necessary, and this should be discussed with the Local Authority.

E1.5 Charges

Approved Inspectors will agree a charge with their client through discussion and the contractual relationship. For a Local Authority, the level of charges is determined by a scheme that each authority is required to set.

E1.6 Building over a sewer
Where works involve the erection, extension or underpinning of a building, they must not be detrimental to the continued maintenance of a drain, sewer or disposal main; as well as the building itself. The building control body must consult with the sewerage undertaker and take account of their views if the sewer is in close proximity or being built over.

E1.7 Requirements of other related legislation
E1.7.1 Local Acts
Some Local Acts control matters relating to particular types of buildings, and these may differ from Building Regulation requirements. For further advice, consult with the building control body or consultant on any additional requirements that may apply.

E1.7.2 Regulatory Reform (Fire Safety) Order
The Regulatory Reform (Fire Safety) Order[125] replaced most fire safety legislation in 2006. It requires any person who has some level of control in premises to reduce the risk from fire and ensure safe escape by reasonable measures. It does not apply to private homes or individual flats but covers all other buildings; including the common areas of blocks of flats, the external walls, and certain fire safety elements within the flat. A building control body will need to send Building Regulation submissions to the fire and rescue authority for comments on any buildings covered by the Order and it is important that relevant information is submitted in good time to avoid delays in approval or certification.

E1.8 Future developments
The Building Safety Act[22] applies major changes to the building control system in England and Wales. The Act is being commenced throughout 2023 and 2024; therefore some of the terminology used in this guidance may change. Where further clarity or information is needed, consult with your building control body.

The Act establishes the Building Safety Regulator in England within the Health and Safety Executive and will oversee the building control regime in England, including the safety and performance of all buildings, advising on the regulations and producing guidance.

It introduces more stringent competence requirements for anyone working with buildings, including responsibilities for duty holders acting as a client, principal designer/designer or principal contractor/contractor for all work subject to Building Regulations.

In addition, buildings classed as 'higher risk' will be under the supervision of the Building Safety Regulator rather than a Local Authority or Approved Inspector. Although outside of the scope of this publication, your building control body should be able to advise the process to follow (if relevant) at the time of submission.

E2 Scotland

E2.1 Introduction
Building control in Scotland is based on the administration by Local Authorities of the Building (Scotland) Regulations 2004[17].

The Building (Scotland) Regulations are made by Scottish Ministers under the powers in Section 1 of The Building (Scotland) Act 2003[126] as amended. The Regulations do not apply retrospectively to existing buildings.

Scottish Ministers, in framing Building Regulations, are required to consult such persons as appear to be representative of the interests concerned.

Regulation 9 (Schedule 5) sets out how the requirements of the regulations can be satisfied:

- By compliance with the relevant guidance set out in the supporting Technical Handbooks[92–93]
- By conforming with provisions which are stated in the Technical Handbooks to be deemed to satisfy the relevant standards
- By any other means which can be shown to satisfy the relevant standards

The designs, materials and methods of construction must therefore be at least to the standards set out in the Technical Handbooks to satisfy regulations.

E2.2 Technical Handbooks

The Technical Domestic and Non-Domestic Handbooks give information on the general application of the regulations and provide:

- A full list of defined terms used in both the Statutory Instruments and Technical Handbooks
- Instructions on the interpretation of references to other publications
- A complete extract of regulations applied, generally covering:
  - Limited life buildings
  - Measurements
  - Fitness and durability of materials and workmanship
  - Building Standards — construction
  - Building Standards — demolition
  - Building Standards — service, fitting or equipment
  - Building Standards — conversion
  - Provision of protective works
  - Clearing of footpaths
  - Securing of unoccupied and partly completed buildings
  - Relaxations

The exemptions defined in the Technical Handbooks are similar to those noted for England and Wales covering the same broad range, subject to exceptions.

E2.3 Schedule 5: Structure

The intention of this Schedule is to ensure that the structure of a building is safe. The relevant criteria for all cases are given in Section 1 of the Technical Handbooks:

- Loadings on the building
- Nature of the ground
- Collapse or deformations
- Stability of the building and other buildings
- Climatic conditions
- Materials
- Structural analysis
- Details of construction

This Schedule also notes that loads used in calculations must allow for possible dynamic, concentrated and peak load effects that may occur.

Schedule 5, Section 1 states that:

> "Every building must be designed and constructed in such a way that the loadings that are liable to act on it will not lead to; the collapse of the whole or part of the building, or deformations which would make the building unfit for its intended use, unsafe, or cause damage to other parts of the building or to fittings or to installed equipment."

Section 1 also states that:

> "Every building must be designed and constructed in such a way that in the event of damage occurring to any part of the structure the extent of any resultant collapse will not be disproportionate to the cause."

E2.4 Further requirements

Further requirements are found in Schedule 5, Sections 3.3–3.6 of the Regulations relating to the preparation of sites and resistance to moisture. These mirror those of England and Wales in many aspects in referring to a number of British Standards and Codes of Practice that satisfy compliance with the substantive requirement of the regulations.

Generally in Scotland, Local Acts were replaced by the Civic Government (Scotland) Act 1982[127] and thus do not pose a difficulty. Other legislation might be relevant in their Scottish context and further advice might be needed.

Partial underpinning, if considered to comply with the relevant standards, will generally be approved.

At the present time there are no Approved Inspectors in Scotland, but Building Warrant applications with structural design that are submitted to Local Authorities in Scotland are to be certified by an Approved Certifier registered under the IStructE's Structural Engineers Registration Ltd (SER) scheme[128]. The Certifying Engineer is responsible for ensuring that all aspects of design satisfy the requirements of the Building (Scotland) Regulations 2004.

E3 Northern Ireland

E3.1 Introduction
The Building Regulations (Northern Ireland) Order 1979[94] confers power on the Department of Finance (NI) to make regulations for:

- All or any of the matters as set out in Schedule 1 of the Order
- Such other matters relating to buildings considered relevant to the purposes of the Order, after due consultations with the Building Regulations Advisory Committee

In the Order 'building' can be defined generally as any permanent or temporary building, and any reference to a building includes a reference to part of a building.

Building Regulations may be expressed in terms of performance, types of material, methods of construction or of ensuring safety (or otherwise). Requirements specified in Building Regulations set reasonable levels to secure the health, safety, welfare and convenience of persons in or about buildings and of others who may be affected by buildings or matters connected with buildings. They also set out to further the conservation of fuel and power, the protection and enhancement of the environment, and promote sustainable development.

Within Northern Ireland the Building Regulations are administered and enforced by district councils and no provision for Approved Inspectors exists (unlike England and Wales).

E3.2 Scope
The Building Regulations (Northern Ireland) 2012[18] requires valid applications and appropriate notices to be made to the district council for the:

- Erection of buildings
- Alteration and extension of buildings
- Provision of services and fittings to buildings
- Material change of use of buildings

The method of application can be either a full plans submission, or building notice application (the notice route only being valid for existing dwellings with certain exclusions).

Building Regulations are expressed in functional format, specifying performance standards to be reasonable and adequate for their purpose. Supporting 'Technical Booklets'[95] are provided to give guidance on demonstrating compliance with the requirements of relevant regulations. However, their use is not mandatory and other solutions can be adopted with suitable justification. Note, that there is a presumption of compliance by following the guidance in a Technical Booklet, although this is not guaranteed and consideration of achieving the functional regulation needs to be given.

E3.2.1 Exemptions
Certain exemptions apply to some building types or certain types of small extension. In addition, certain buildings of statutory undertakers or other specified bodies are exempt. Your building control body can give you further guidance on whether a building is likely to be exempt.

E3.3 How to ensure compliance
The Building Regulations are the mandatory requirements for the design and construction of relevant buildings. The method chosen to demonstrate compliance is open for choice by the applicant or their agent.

For assessment purposes, district councils are likely to benchmark alternative or innovative proposals against the guidance within the Technical Handbooks. Note, that these documents are not regulations in themselves and therefore do not set mandatory requirements. They do indicate methods and standards of construction, which if followed, give a presumption of compliance. However, this is not guaranteed and due consideration of achieving the functional regulation needs to be given.

E3.4 Regulations as they apply to foundations

'Building work' means the erection of a building, the structural alteration or extension of a building (including work in connection with the making of a material change of use) or the provision of a service or fitting.

There is no specific definition for work relating to the underpinning of a building within the Northern Ireland regulations; such work will tend to be considered under Regulation 7 in Part A 'Applications to alterations and extensions'.

E3.4.1 Part D: Structure

Building work shall be carried out so that it complies with the relevant requirements of the regulations. The requirements for foundations and other structural elements of a building are contained in The Building Regulations (Northern Ireland) 2012[18].

Regulation 30 Stability in Part D:

30. A building shall be designed and constructed so that the combined dead, imposed and wind loads are sustained and transmitted to the ground, taking into account the nature of the ground:
    (a) safely, and
    (b) without impairing the safety of any part of another building.

Guidance within Section 2 of Technical Booklet D[96] provides references to standards that can be used to demonstrate compliance with the regulation.

There is also guidance on elements of construction for certain residential buildings or other 'small buildings' and Section 4E of Technical Booklet D relates to strip foundations of plain concrete.

E3.4.2 Further requirements

Additional requirements need to be considered for foundations and underpinning of new buildings, altering or extending existing buildings, and to buildings undergoing a material change of use. Specialist advice is recommended to determine if any of the other functional regulations will be affected or apply when undertaking works, and consultation with the relevant district council is also recommended.

Of note are some of the functional regulations in relation to Part C, 'Site preparation and resistance to contaminants and moisture', including:

- Regulation 26: Site preparation and resistance to contaminants
- Regulation 27: Subsoil drainage
- Regulation 28: Resistance to moisture and weather

Guidance in support of Part C, and Regulations 26–28 can be found in Technical Booklet C[129].

Appendix F Underpinning guarantee

This appendix contains a specimen underpinning guarantee.

The Association of Specialist Underpinning Contractors (ASUC) is the UK trade association for underpinning and subsidence repairs. Warranties have strengthened considerably in recent years, and ASUC members are able to give a 10-year Latent Defects Warranty for works carried out.

Latent Defects Insurance is similar to the NHBC/Premier Guarantee/LABC Warranty product used for new builds — with the added benefit that cover commences from day one after practical completion of the project.

A further additional benefit is that it is a direct contract between the Insurer and the client, and therefore not dependent on the contractor still trading (for whatever reason) or an Insurer taking the contractor guarantee and underwriting a loss, should the contractor cease to trade, leaving the client having to prove negligence to make a claim. Known as an 'indemnity policy' the insurance reacts to a claim irrespective of proof of negligence, merely the need to prove a claim.

As with all insurance policies, terms and conditions apply, together with an excess or minimum claim value. At the point of sale, any property that has been subjected to subsidence works, whether that be structural repairs and/or underpinning is notifiable, and the conveyancing Solicitor will require proof of the works and a copy of the warranty. The warranty is attached to the property, not the individual/client and is therefore transferable without any further actions.

Ten Year Defects Guarantee

ASUC FULL CONTRACTING MEMBER COMPANY:
XXXXXXXXXXXX. REGISTRATION NUMBER FOR ASUC XXXXXXX INSERT MEMBER CO LOGO IF DESIRED.

Contract Address:

Client:

Contract Address:

Contract sum (and limit of Guarantee sum)

Description of works completed.

Completion date

Guarantee expiry date

ASUC member trading address

Phone

E mail

This guarantee is protected by an Insurance backed Warranty issued by the ASUC insurers under the following policy no:

In the event the ASUC member has ceased to trade for whatever reason the insurer will fulfil their duties under the terms of the policy which is attached with this certificate of Guarantee. ASUC have no liability in respect of this Guarantee.

Welcome to Your Defects Guarantee.

You have the benefit of this Guarantee as the Contracted Works undertaken at Your Premises have been completed by a member of ASUC. ASUC is an independent trade association promoting professional and technical competence. Its membership comprises leading specialist contractors and associated companies within the subsidence repairs, underpinning, retrofit basements and engineered foundation solutions. Further details are available at www.asuc.org.uk.

Although it is unlikely that there will be problems with Your Premises, the Guarantee gives you the comfort of knowing that particular types of problem which may arise in the first ten years following completion of the Contracted Works will be corrected.

You need to read these terms and conditions, Your Certificate of Guarantee and any endorsements attached to them, to make sure You know what is covered by the Guarantee.

ASUC accept no liability for the costs or any remedial works. The Guarantee is issued by the ASUC member as shown on the certificate of Guarantee and the contract is between the ASUC member listed and you the client.

ASUC have a complaints procedure in place in the event that you do not think the Member company have carried out their duties as contracted. Clients can make a complaint in writing with supporting evidence to the ASUC office.

ASUC
Unit 34 Base Bordon Innovation Centre Broxhead House, 60 Barbados Rd, Bordon,
GU35 0FX.
T 01420 471613
E admin@asuc.org.uk

The outcome of any complaint to ASUC as carried out by the Executive Board is final with no further appeal procedure.

HOW TO MAKE A CLAIM
We know how difficult and stressful it can be if things go wrong with Your Premises and we aim to work with you to ensure that we can get You back to normality as quickly as possible. In order to make a claim contact directly the ASUC member who carried out the works with a copy of this guarantee document.

WHAT IS COVERED
Physical loss, destruction or damage to the Premises or a condition requiring immediate remedial action to prevent imminent structural instability to the Premises, for a value up to the contract sum caused by a Defect and which is first discovered during the Period of Guarantee

Any defect in the design, where the design is carried out by the contractor, specification, workmanship, materials, or components of the Contracted Works up to the contract sum specified on the Certificate.

EXAMPLES OF CONTRACTED WORKS
Retrofit Basement works including Waterproofing works if part of the contract.
Subsidence Repair works including Structural repairs.
Underpinning
Foundation and Piling works.

YOUR RESPONSIBILITIES

1. On discovery of any occurrence or circumstance that is likely to give rise to a claim, it is a Condition Precedent that the Client should, as soon as possible please contact the ASUC member
2. Once You have notified the ASUC member of a potential claim, it is a Condition Precedent that You:
   (a) send evidence (including photographs and reports if appropriate) to the ASUC member confirming that there has been damage; and
   (b) take all responsible steps to prevent further loss or damage; and
   (c) allow The ASUC member to have all reasonable access to the Premises during normal working hours; and
   (d) obtain permission to access neighbouring land and obtain any other permission needed to allow investigations and work; and
   (e) To the extent that this is in Your possession or control, provide the ASUC member with all information and documentation it may request which is necessary in order to investigate the claim
   (f) Supply the ASUC member with a copy of this guarantee document

CONDITIONS
No additional costs can be considered such as alternative accommodation, indexation or consequential damage or fees incurred by guarantee holder.

The Company shall not be responsible under this Guarantee if:

(i) the Customer shall not have paid in full all invoices for work done or goods supplied by the Company within the time agreed:
(ii) the work done, or goods supplied have been subject to misuse, neglect, negligence, or accident:
(iii) the work had been improperly maintained:

No claim for breach of warranty shall entitle the Customer to cancel any contract for work by the Company. The warranty contained in this Clause shall be deemed to take the place of any other guarantee, condition, warranty, or other liability implied by law (including implied terms as to merchantability, fitness for a particular purpose, correspondence with description or otherwise) all of which are hereby expressly excluded to the extent to which the law permits.

EXCLUSIONS

1 FLOODING AND WATER TABLE
Loss or damage resulting from or in connection with flooding or failure of flood prevention / defence measures, however caused, or from a change in the water table level. This exclusion shall not however apply to any seasonal change in the water table level.

2. HUMIDITY
Loss or damage caused by, in connection with or consequent upon humidity, normal dampness or condensation in Premises that is not the direct result of Damage.

3. INDIRECT LOSS
Unless expressly provided for in this Guarantee, economic or financial loss of any description (including costs arising from inconvenience or distress, loss of enjoyment, loss of use, reduction in value of the Premises, loss of income or business opportunity) arising either directly or indirectly as a result of the events or circumstances that led to Your claim.

4. MAINTENANCE AND USE
Inadequate maintenance of the Contracted Works or the imposition of any load greater than that for which the Contracted Works was designed or the use of a Premises for any purpose other than that for which it was designed, unless (in each case) the ASUC member has been informed in writing, and the Guarantee has been endorsed.

5. PERSONAL INJURY
Any costs, losses, expenses or damages for death, bodily injury, disease, illness, or injury to mental health, however caused.

6. RADIOACTIVE CONTAMINATION, CHEMICAL, BIOLOGICAL, BIO-CHEMICAL AND ELECTROMAGNETIC WEAPONS
Any loss or destruction of or damage to any property, any loss or expense whatsoever, any consequential loss or any legal liability directly or indirectly caused by or contributed to by, or arising from:
(a) (i) ionising radiations or contamination by radioactivity from nuclear fuel or from nuclear waste from the combustion of nuclear fuel;
(ii) the radioactive, toxic, explosive or other hazardous or contaminating properties of any nuclear installation, reactor or other nuclear assembly or nuclear component thereof
(b) the use of any weapon or device:
(i) dispersing radioactive material and / or ionising radiation; or
(ii) using atomic or nuclear fission and / or fusion or other like reaction
(c) the radioactive, toxic, explosive or other hazardous or contaminating properties of any radioactive matter but this will not apply in respect of radioactive isotopes (other than nuclear fuel or nuclear waste) used for the purposes for which they were intended;
(d) any chemical, biological, bio-chemical, or electromagnetic weapon.

7. SEEPAGE
Loss or damage caused by seepage of water into the Premises below ground floor slab level. Except where the contracted works specifically includes Waterproofing works in Basement contracts

8. SETTLEMENT AND OTHER GROUND MOVEMENT
Loss or damage caused by, associated with or in connection with any movement of land or settlement of bedding down of the Foundation Works, unless the loss or damage is caused by a Defect.

9. SONIC BANGS
Loss or damage occasioned by or connected with pressure waves caused by aircraft or other aerial devices travelling at sonic or supersonic speeds.

10. SPECIAL PERILS
Loss or damage which would be covered under a standard buildings insurance policy (whether or not You have in fact purchased such a policy). For example, this means that we do not provide cover for loss or damage caused or contributed to by any of the following:
(a) fire, lightning, or explosion
(b) earthquake
(c) flood
(d) storm
(e) aircraft or other flying objects, or articles falling from them
(f) water, oil, or any other liquids leaking or bursting from tanks, pipes, heating systems or other equipment
(g) malicious acts (such as vandalism or criminal damage)
(h) theft or attempted theft; or
(i) accidental damage

11. SUBSIDENCE
Loss or damage caused by or consequent upon subsidence, heave, or landslip, except where such loss or damage is as a direct result of a Defect.

12. TERRORISM
Any loss, damage, liability, cost, or expense of whatsoever nature directly or indirectly caused by, resulting from, arising out of or in connection with an act of terrorism. For the purposes of this exclusion, an act of terrorism shall mean an act involving the use of force or violence and / or the threat thereof that is committed for political, religious, or ideological purposes and with the intention to influence any government and / or to put the public, or any section of the public, in fear. The perpetrators of an act of terrorism can be acting alone or on behalf of, or in connection with, any organisation(s) or government(s).

13. THIRD PARTY DESIGN
Loss or damage caused by defective design of the Contracted Works where the Works were designed by a party other than the Contractor.

14. TREES
Loss or damage due to, associated with or arising from:
(a) trees planted after the completion of the Contracted Works that cause damage to the Premises; or
(b) the failure of the client to comply with all recommendations stipulated by the Structural Engineer or the ASUC member appointed in respect of the Contracted Works in so far as such recommendations are applicable to the maintenance and or removal of trees at the Premises or adjoining properties

15. VERMIN
Loss or damage caused by, connected with or consequent upon the actions of rodents, vermin, or insect infestation.

16. WAR RISKS
Any consequence whatsoever which is the direct or indirect result of any of the following, or anything connected with any of the following, whether or not such consequence has been contributed to by any other cause or event:
(a) war, invasion, act of a foreign enemy, hostilities or a warlike operation or operations (whether war be declared or not), civil war, rebellion, revolution, insurrection, civil commotion assuming the proportions of or amounting to an uprising, military or usurped power, mutiny or military uprising, martial law; and
(b) nationalisation, confiscation, requisition, seizure, damage, or destruction by or by order of any government or any local or public authority; and
(c) any action taken in controlling, preventing, suppressing or in any way relating to (a) and / or (b) above

17. WEAR AND TEAR
Loss or damage caused by or associated with any wear and tear, normal dampness, condensation, thermal movement, movement between different types of materials or shrinkage or normal deterioration (whether caused by neglect or otherwise).

18. WILFUL ACTS
Any wilful neglect or criminal act of the Policyholder or any other person.

19. MOULD
Loss or damage arising out of or in connection with any bacteria, yeasts, mildew, viruses, fungi, mould or their spores, mycotoxins or other metabolic products in a Premises that is not the direct result of a defect.

20. THIRD PARTY RIGHTS
A person who is not a party to this Policy has no right under the Contracts (Rights of Third Parties) Act 1999 to enforce any term of this Policy, but this does not affect any right or remedy of a third party which exists or is available apart from that Act.

Signed on behalf of the ASUC member.

Position:

Company:

Date:

ASUC TEN YEAR GUARANTEE WITH IBG POLICY REV 3 RDW/ASUC 26/04/2023

References

1 The Institution of Structural Engineers. *Subsidence of low-rise buildings: A guide for professionals and property owners.* London: SETO, 1994.

2 The Institution of Structural Engineers. *Subsidence of low-rise buildings: A guide for professionals and property owners* (2nd edition). London: SETO, 2000.

3 Hancock M. *The subsidence hangover of 2018.* Available at: www.geplus.co.uk/features/the-subsidence-hangover-of-2018-14-08-2018 [Accessed: October 2023].

4 Davies Group Ltd. *Will a subsidence surge happen this summer?* Available at: https://davies-group.com/knowledge/will-a-subsidence-surge-happen-this-summer [Accessed: October 2023].

5 Financial Ombudsman Service. *Subsidence and other types of ground movement.* Available at: www.financial-ombudsman.org.uk/businesses/complaints-deal/insurance/home-buildings-insurance/subsidence-types-ground-movement [Accessed: October 2023].

6 Driscoll R. *DG 251: Assessment of damage in low-rise buildings, with particular reference to progressive foundation movement. Revised 1995.* Watford: Building Research Establishment, 1995.

7 Intergovernmental Panel on Climate Change (IPCC). *Special Report: Global warming of 1.5°C – Glossary.* Available at: www.ipcc.ch/sr15/chapter/glossary/ [Accessed: October 2023].

8 Building Cost Information Service (BCIS). *Unlocking data*. Available at: https://bcis.co.uk/ [Accessed: October 2023].

9 Investopedia. *What is subrogation?* Available at: www.investopedia.com/terms/s/subrogation.asp#:~:text=Subrogation%20is%20a%20term%20describing,the%20insured%20for%20the%20loss [Accessed: October 2023].

10 RICS. *Home survey standard.* Available from: www.rics.org/profession-standards/rics-standards-and-guidance/sector-standards/building-surveying-standards/home-surveys/home-survey-standards [Accessed: October 2023].

11 Financial Conduct Authority. *Insurance Conduct of Business Sourcebook (ICOBS).* Available at: www.handbook.fca.org.uk/handbook/ICOBS [Accessed: October 2023].

12 HM Government. *Equality Act 2010*. Available at: www.legislation.gov.uk/ukpga/2010/15 [Accessed: October 2023].

13 Financial Conduct Authority. *Guidance for firms on the fair treatment of vulnerable customers*. Available at: www.fca.org.uk/publications/finalised-guidance/guidance-firms-fair-treatment-vulnerable-customers [Accessed: October 2023].

14 HM Government. *Party Wall etc. Act 1996.* Available at: www.legislation.gov.uk/ukpga/1996/40 [Accessed: October 2023].

15 HM Government. *The Construction (Design and Management) Regulations 2015.* Available at: www.legislation.gov.uk/uksi/2015/51 [Accessed: October 2023].

16 HM Government. *The Building Regulations 2010.* Available at: www.legislation.gov.uk/uksi/2010/2214 [Accessed: October 2023].

17 HM Government. *The Building (Scotland) Regulations 2004.* Available at: www.legislation.gov.uk/ssi/2004/406 [Accessed: October 2023].

18 HM Government. *The Building Regulations (Northern Ireland) 2012.* Available at: www.legislation.gov.uk/nisr/2012/192 [Accessed: October 2023].

19 HM Government. *Housing Act 1985*. Available at: www.legislation.gov.uk/ukpga/1985/68 [Accessed: October 2023].

20 HM Government. *Housing Act 2004.* Available at: www.legislation.gov.uk/ukpga/2004/34 [Accessed: October 2023].

21 HM Government. *Housing Grants, Construction and Regeneration Act 1996.* Available at: www.legislation.gov.uk/ukpga/1996/53 [Accessed: October 2023].

22 HM Government. *Building Safety Act 2022.* Available at: www.legislation.gov.uk/ukpga/2022/30/contents/enacted [Accessed: October 2023].

23 HM Government. *Financial Services and Markets Act 2000.* Available at: www.legislation.gov.uk/ukpga/2000/8/contents [Accessed: October 2023].

24 Association of British Insurers. *Domestic subsidence/heave/landslip "change of insurer" claims agreement.* Available at: www.abi.org.uk/globalassets/files/subject/public/home-insurance/abi-domestic-subsidence-claims-agreement-and-guidelines-dec-2017.pdf [Accessed: October 2023].

25 Association of British Insurers. *Domestic subsidence tree root claims agreement third party liability.* Available at: www.abi.org.uk/globalassets/files/subject/public/home-insurance/abi-domestic-subsidence-tree-root-claims-agreement-may-2017.pdf [Accessed: October 2023].
26 HM Government. *Consumer Insurance (Disclosure and Representations) Act 2012.* Available at: www.legislation.gov.uk/ukpga/2012/6/contents/enacted [Accessed: October 2023].
27 HM Government. *Fraud Act 2006.* Available at: www.legislation.gov.uk/ukpga/2006/35/contents [Accessed: October 2023].
28 Association of British Insurers. *Calculating your rebuild cost.* Available at: www.abi.org.uk/products-and-issues/choosing-the-right-insurance/home-insurance/buildings-insurance/calculating-your-rebuild-cost/ [Accessed: October 2023].
29 HM Government. *Climate Change Act 2008.* Available at: www.legislation.gov.uk/ukpga/2008/27/contents [Accessed: October 2023].
30 British Standards Institution. *BS 1377-2:2022. Methods of test for soils for civil engineering purposes — Classification tests and determination of geotechnical properties.* London: BSI, 2022.
31 British Standards Institution. *BS EN ISO 17892-5:2017. Geotechnical investigation and testing. Laboratory testing of soil — Incremental loading oedometer test*. London: BSI, 2017.
32 NHBC. *NHBC Standards: 4.2 Building near trees.* Available at: https://nhbc-standards.co.uk/4-foundations/4-2-building-near-trees/ [Accessed: October 2023].
33 Tomlinson MJ. *Foundation design and construction* (7th edition). Harlow: Prentice Hall, 2001.
34 Stantec. *Cavities Database*. Available at: www.stantec.com/uk/services/cavities-database [Accessed: October 2023].
35 CIRIA. *Ground-borne vibrations arising from piling (TN142D).* London: CIRIA, 1992.
36 British Standards Institution. *BS 5228-2:2009+A1:2014. Code of practice for noise and vibration control on construction and open sites — Vibration.* London: BSI, 2009.
37 British Standards Institution. *BS 6472-1:2008. Guide to evaluation of human exposure to vibration in buildings — Vibration sources other than blasting.* London: BSI, 2008.
38 BRE Construction Division. *Concrete in aggressive ground (SD1).* Watford: BRE, 2005.
39 British Standards Institution. *BS 1377-3:2018+A1:2021. Methods of test for soils for civil engineering purposes — Chemical and electro-chemical testing.* London: BSI, 2021.
40 HM Government. *Thaumasite Expert Group Report: Review after three years experience.* Available at: https://assets.publishing.service.gov.uk/government/uploads/system/uploads/attachment_data/file/429277/thaumasite_three_year_review.pdf [Accessed: October 2023].
41 British Standards Institution. *BS 5837:2012. Trees in relation to design, demolition and construction. Recommendations.* London: BSI, 2012.
42 RICS. *The mundic problem.* Available at: www.rics.org/profession-standards/rics-standards-and-guidance/sector-standards/building-surveying-standards/the-mundic-problem [Accessed: October 2023].
43 British Geological Survey. *BGS Geology Viewer.* Available at: www.bgs.ac.uk/map-viewers/bgs-geology-viewer/ [Accessed: October 2023].
44 The Planning Portal. *Transforming Planning and Building*. Available at: www.planningportal.co.uk/ [Accessed: October 2023].
45 Bonshor R, Bonshor L and Sadgrove R. *Cracking in buildings (BR 292 2e 2016)*. Watford: BRE, 2016.
46 Building Research Establishment. *Simple measuring and monitoring of movement in low-rise buildings. Part 1: Cracks (DG 343)*. Watford: BRE, 1989.
47 Longworth I. *Simple measuring and monitoring of movement in low-rise buildings. Part 2: Settlement, heave and out-of-plumb (DG 344).* Watford: BRE, 1989.
48 Building Research Establishment. *Monitoring building and ground movement by precise levelling (DG 386).* Watford: BRE, 1993.
49 Building Research Establishment. *Site investigation for low-rise buildings: direct investigations (DG 411).* Watford: BRE, 1995.
50 British Standards Institution. *BS 1377-9:1990. Methods for test for soils for civil engineering purposes — In-situ tests.* London: BSI, 1990.
51 Building Research Establishment. *Site investigation for low-rise buildings: trial pits (DG 381).* Watford: BRE, 1993.
52 British Standards Institution. *BS 5930:2015+A1:2020. Code of practice for ground investigations.* London: BSI, 2020.
53 Crilly MS and Chandler RJ. *A method of determining the state of desiccation in clay soils.* Watford: BRE, 1993.
54 Crilly MS. *Desiccation in clay soils (DG 412).* Watford: BRE, 1996.

55 Building Research Establishment. *Low-rise buildings on shrinkable clay soils (DG 240).* Watford: BRE, 1993.
56 Trees & Design Action Group. *First steps in valuing trees and green infrastructure.* Available at: www.tdag.org.uk/first-steps-in-valuing.html [Accessed: October 2023].
57 East Malling Research and University of Cambridge. *Controlling water use of trees to alleviate subsidence risk* (Horticulture LINK Project 212). Available at: www.bre.co.uk/filelibrary/pdf/projects/HortLink_Project_Final_Report_(2004).pdf [Accessed: October 2023].
58 Hipps NA and Atkinson CJ. *Pilot study to determine the feasibility of using existing data to determine the impacts of tree pruning on subsidence incidents on swelling clay soils* (Beyond Hortlink 212). Available at: www.subsidenceforum.org.uk/pdf_word/Full%20report%20V10%20(secured).pdf [Accessed: October 2023].
59 Cutler DF, Gasson PE and Farmer MC. Wind-blown tree survey: analysis of results. *Arboricultural Journal*. 1990;14(3): https://doi.org/10.1080/03071375.1990.9746846.
60 Hipps NA, Atkinson CJ and Griffiths H. *Pruning trees to reduce water use: summaries of research; conclusions and recommendations (BRE Information Paper IP7/06).* Watford: BRE Press, 2006
61 London Tree Officers Association. *The risk limitation strategy for tree root claims.* (3rd edition). London: LTOA, 2008. Available at: www.ltoa.org.uk/documents-1/risk-limitation-strategy-for-tree-root-claims/126-the-risk-limitation-strategy-for-tree-root-claims [Accessed: October 2023].
62 London Tree Officers Association. *JMP Tree Pruning Criteria.* London: LTOA, 2010. Available at: https://www.ltoa.org.uk/documents-1/joint-mitigation-protocol/132-jmp-tree-pruning-criteria [Accessed: October 2023].
63 Hirons AD and Thomas PA. *Applied tree biology.* Chichester: Wiley-Blackwell, 2018.
64 Cutler DF and Richardson IBK. *Tree roots and buildings* (2nd edition). Harlow: Longman, 1989.
65 HM Government. *Town and Country Planning Act 1990.* Available at: www.legislation.gov.uk/ukpga/1990/8/contents [Accessed: October 2023].
66 HM Government. *The Town and Country Planning (Tree Preservation)(England) Regulations 2012.* Available at: www.legislation.gov.uk/uksi/2012/605/contents/made [Accessed: October 2023].
67 London Tree Officers Association. *Capital Asset Value for Amenity Trees (CAVAT).* Available at: www.ltoa.org.uk/documents-1/capital-asset-value-for-amenity-trees-cavat [Accessed: October 2023].
68 Association of British Insurers. *Protecting your home from subsidence damage.* Available at: www.abi.org.uk/globalassets/sitecore/files/documents/publications/public/migrated/home/protecting-your-home-from-subsidence-damage.pdf [Accessed: October 2023].
69 LABC Warranty. *Reducing the effects of ground heave following tree removal.* Available at: www.labcwarranty.co.uk/technical-blog/reducing-the-effects-of-ground-heave-following-tree-removal [Accessed: October 2023].
70 HM Government. *Forestry Act 1967.* Available at: https://www.legislation.gov.uk/ukpga/1967/10/contents [Accessed: October 2023].
71 London Tree Officers Association. *Joint Mitigation Protocol.* Available at: www.ltoa.org.uk/resources/joint-mitigation-protocol [Accessed: October 2023].
72 HM Government. *Access to Neighbouring Land Act 1992.* Available at: https://www.legislation.gov.uk/ukpga/1992/23/contents [Accessed: October 2023].
73 HM Government. *Defective Premises Act 1972.* Available at: https://www.legislation.gov.uk/ukpga/1972/35/contents [Accessed: October 2023].
74 Global Association of Risk Professionals. *The 4 Main Drivers of Transition Risk, and Why the Risks Are Increasing.* Available at: www.garp.org/risk-intelligence/sustainability-climate/4-main-drivers-of-transition-risk-220412 [Accessed: October 2023].
75 Buzzi O, Fityus S and Sloan SW. Use of expanding geopolymer resin to remediate expansive soil foundations. *Canadian Geotechnical Journal.* 2010. Available from: https://cdnsciencepub.com/doi/10.1139/T09-132 [Accessed: October 2023].
76 British Standards Institution. *BS EN 1997-1:2004+A1:2013: Eurocode 7. Geotechnical design — General rules.* London: BSI, 2004.
77 British Standards Institution. *BS 8004:2015+A1:2020: Code of practice for foundations.* London: BSI: 2020.
78 British Standards Institution. *BS 8500-1:2015+A2:2019: Concrete. Complementary British Standard to BS EN 206 — Method of specifying and guidance for the specifier.* London: BSI, 2020.
79 ASUC. *Guidelines on safe and efficient underpinning and mini piling operations.* (3rd edition). Available at: https://asuc.associationhouse.org.uk/default.php?cmd=210&doc_category=97 [Accessed: October 2023].
80 Institution of Civil Engineers. *ICE specification for piling and embedded retaining walls.* (3rd edition). London: ICE Publishing, 2016.

81 British Standards Institution. *BS 10175:2011+A2:2017: Investigation of potentially contaminated sites. Code of practice*. London: BSI, 2017.
82 British Standards Institution. *BS EN 1997-2:2007: Eurocode 7. Geotechnical design — Ground investigation and testing.* London: BSI: 2007.
83 British Standards Institution. *NA to BS EN 1997-2:2007: UK National Annex to Eurocode 7. Geotechnical design — Ground investigation and testing.* London: BSI, 2009.
84 Joint Contract Tribunal (JCT). *Minor works building contract 2016.* London: Sweet & Maxwell Ltd, 2016.
85 HM Government. *Health and Safety at Work, etc. Act 1974.* Available at: www.legislation.gov.uk/ukpga/1974/37 [Accessed: October 2023].
86 HM Government. *The Provision and Use of Work Equipment Regulations 1998.* Available at: www.legislation.gov.uk/uksi/1998/2306 [Accessed: October 2023].
87 HM Government. *The Lifting Operations and Lifting Equipment Regulations 1998.* Available at: www.legislation.gov.uk/uksi/1998/2307 [Accessed: October 2023].
88 HM Government. *The Management of Health and Safety at Work Regulations 1999.* Available at: www.legislation.gov.uk/uksi/1999/3242/contents/made [Accessed: October 2023].
89 HM Government. *The Reporting of Injuries, Diseases and Dangerous Occurrences Regulations 2013.* Available at: www.legislation.gov.uk/uksi/2013/1471 [Accessed: October 2023].
90 HM Government. *The Control of Substances Hazardous to Health Regulations 2002.* Available at: www.legislation.gov.uk/uksi/2002/2677 [Accessed: October 2023].
91 Construction Industry Council Approved Inspectors Register (CICAIR). *Overseeing continuous improvement in the quality and standards of Approved Inspectors through proportionate, targeted and effective building control regulation.* Available at: www.cicair.org.uk [Accessed: October 2023].
92 Scottish Government. *Building standards technical handbook 2022: domestic.* Available at: www.gov.scot/publications/building-standards-technical-handbook-2022-domestic [Accessed: October 2023].
93 Scottish Government. *Building standards technical handbook 2022: non-domestic.* Available at: www.gov.scot/publications/building-standards-technical-handbook-2022-non-domestic [Accessed: October 2023].
94 HM Government. *Building Regulations (Northern Ireland) Order 1979.* Available at: www.legislation.gov.uk/nisi/1979/1709 [Accessed: October 2023].
95 Department of Finance. *Building Regulations Technical Booklets.* Available at: www.finance-ni.gov.uk/articles/building-regulations-technical-booklets [Accessed: October 2023].
96 Department of Finance. *The Building Regulations (Northern Ireland) 2012: Guidance. Technical Booklet D: Structure*. Available at: www.finance-ni.gov.uk/publications/technical-booklet-d [Accessed: October 2023].
97 HM Government. *Find out if a property is affected by coal mining.* Available at: www.gov.uk/check-if-property-is-affected-by-coal-mining [Accessed: October 2023].
98 Parry DN and Chiverrell CP. [eds.] *Abandoned mine workings manual* (CIRIA Publication C758D). London: CIRIA, 2019.
99 Institution of Civil Engineers. *ICE manual of geotechnical engineering: Volume I.* London: ICE Publishing, 2012.
100 Institution of Civil Engineers. *ICE manual of geotechnical engineering: Volume II.* London: ICE Publishing, 2012.
101 Coppin NJ and Richards IG. [eds.] *Use of vegetation in civil engineering.* London: CIRIA, 1990.
102 The Institution of Structural Engineers. *Guide to surveys and inspections of buildings and associated structures* (2nd edition). London: IStructE Ltd, 2008.
103 Department for Levelling Up, Housing & Communities. *The Party Wall etc Act 1996: explanatory booklet.* Available at: www.gov.uk/government/publications/preventing-and-resolving-disputes-in-relation-to-party-walls/the-party-wall-etc-act-1996-explanatory-booklet [Accessed: October 2023].
104 Dines HG. *The metalliferous mining region of South West England* (2 volumes). London: HMSO, 1956.
105 British Geological Survey. *Arup Review of Mining Instability in Great Britain.* Available at: https://metadata.bgs.ac.uk/geonetwork/srv/eng/catalog.search#/metadata/9df8df51-63f2-37a8-e044-0003ba9b0d98 [Accessed: October 2023].
106 Donnelly L. Coal mining subsidence in the UK. *Engineering Geology Special Publications.* The Geological Society. 2020. Available at: www.lyellcollection.org/doi/10.1144/EGSP29.11 [Accessed: October 2023].
107 McTrusty JW. Control of mining subsidence. *Colliery Engineering*. 1959.
108 Whetton JT. A general survey of the ground movement problem. *The Engineer.* 1957.
109 Donnelly L. Mining-induced fault reactivation in the UK. *Engineering Geology Special Publications.* The Geological Society. 2020. Available at: www.lyellcollection.org/doi/10.1144/1470-9236/05-015 [Accessed: October 2023].

110 Beevers C and Wardell K. Recent research in mining subsidence. *Transactions of the Institution of Mining and Metallurgy*. 1954.
111 Wardell K. Some observations on the relationship between time and mining subsidence. *Transactions of the Institution of Mining and Metallurgy.* 1953.
112 Orchard RJ. Recent developments in predicting the amplitude of mining subsidence. *Journal of the Royal Institution of Chartered Surveyors.* 1954.
113 Thorburn S and Reid WG. Incipient failure and demolition of two-storey dwellings due to large ground movements. In Geddes JD. [ed.] *Large ground movements and structures.* London: Pentech Press, 1978.
114 Cameron DWC. Menace of present day subsidence due to ancient mineral operations. *Journal of the Royal Institution of Chartered Surveyors* (Scottish Supplement). 1956.
115 Dean JW. Old mine shafts and their hazards. *The Mining Engineer.* 1967.
116 Price DG. Engineering geology in the urban environment. *Quarterly Journal of Engineering Geology*. 1971. Available at: https://doi.org/10.1144/GSL.QJEG.1971.004.03.03 [Accessed: October 2023].
117 Price DG, Malkin AB and Knill JL. Foundations of multi-storey blocks on coal measures with special reference to old mine workings. *Quarterly Journal of Engineering Geology.* 1967. Available at: https://doi.org/10.1144/GSL.QJEG.1969.001.04.03 [Accessed: October 2023].
118 HM Government. *Coal Mining Subsidence Act 1991.* Available at: www.legislation.gov.uk/ukpga/1991/45 [Accessed: October 2023].
119 HM Government. *Building Act 1984.* Available at: www.legislation.gov.uk/ukpga/1984/55 [Accessed: October 2023].
120 HM Government. *The Construction Products (Amendment etc.) (EU Exit) Regulations 2020.* Available at: www.legislation.gov.uk/uksi/2020/1359/contents [Accessed: October 2023].
121 HM Government. *The Building Regulations 2010. Approved Document A: Structure.* Available at: www.gov.uk/government/publications/structure-approved-document-a [Accessed: October 2023].
122 Welsh Government. *The Building Regulations 2010. Approved Document A: Structure.* Available at: https://gov.wales/building-regulations-guidance-part-structural-safety [Accessed: October 2023].
123 HM Government. *Approved Documents.* Available at: www.gov.uk/government/collections/approved-documents [Accessed: October 2023].
124 Welsh Government. *Building Regulations: Approved documents*. Available at: https://gov.wales/building-regulations-approved-documents [Accessed: October 2023].
125 HM Government. *The Regulatory Reform (Fire Safety) Order 2005.* Available at: https://www.legislation.gov.uk/uksi/2005/1541 [Accessed: October 2023].
126 HM Government. *Building (Scotland) Act 2003.* Available at: https://www.legislation.gov.uk/asp/2003/8 [Accessed: October 2023].
127 HM Government. *Civic Government (Scotland) Act 1982.* Available at: www.legislation.gov.uk/ukpga/1982/45 [Accessed: October 2023].
128 Structural Engineers Registration Ltd. *We provide and manage registration schemes for the certification of the design of building structures*. Available at: ser-ltd.com [Accessed: October 2023].
129 Department of Finance. *The Building Regulations (Northern Ireland) 2012: Guidance. Technical Booklet C: Site preparation and resistance to contaminants and moisture.* Available at: www.finance-ni.gov.uk/publications/technical-booklet-c [Accessed: October 2023].